Edexcel
GCSE Statistics

Gill Dyer

Jane Dyer

David Kent

Keith Pledger

Brian Roadnight

Gordon Skipworth

Examination practice papers by
Chris Frost and Gill Dyer

www.heinemann.co.uk

✓ Free online support
✓ Useful weblinks
✓ 24 hour online ordering

01865 888058

Heinemann
Inspiring generations

Heinemann Education Publishers
Halley Court, Jordan Hill, Oxford OX2 8EJ
Part of Harcourt Education

Heinemann is the registered trademark of
Harcourt Education Limited

© Gill Dyer, Jane Dyer, Brian Roadnight, Gordon Skipworth, David Kent, 2003

First published 2003

08 07 06 05 04
10 9 8 7 6 5 4

British Library Cataloguing in Publication Data is available
from the British Library on request.

ISBN 0 435 53312 6

Edited by Anne Russell
Designed and produced by Tech-Set Ltd, Gateshead, Tyne and Wear
Typeset by Tech-Set Ltd

Original illustrations © Harcourt Education Limited, 2003

Illustrated by Tech-Set Ltd
Cover design by GD Associates
Printed by Mateu

Acknowledgements
Every effort has been made to contact copyright holders of material reproduced in this
book. Any omissions will be rectified in subsequent printings if notice is given to the
publishers.

Tel: 01865 888058 email: info.he@heinemann.co.uk

Contents

Note: red lines in the left hand margin denote work for the Higher Tier.

1 Introduction

1.1 The basis of statistics

Statistics is used when you want to make sense of large amounts of information or data. One part of statistics is concerned with the **description** of data so that ordinary people can understand it. We can do this by representing the data graphically. You will meet a variety of different graphical methods in this book. These are known as **descriptive statistics**. Another way of making large amount of data easily understood is by **summarising** the data. Data may be summarised by applying mathematical processes and finding things like averages and ranges as you may have done in Key Stage 3. These are known as **summary statistics**. Chapters 3 and 4 of this book are about descriptive statistics. Chapter 5 deals with summary statistics

A widely used statistical process is concerned with making inferences about a large amount of data (known by statisticians as the **population**) by examining a few members of it – a process known as **sampling**. This is known as **statistical analysis**. The basis of all statistical analysis is the obtaining of good, reliable data. The data needs to be both accurate and in a useable form, so that any sample is truly representative of the whole population, and may be used to draw correct and accurate inferences about the population. Good statistical analysis is only possible if the data has been collected properly. Chapter 2 is about data collection.

The production of both descriptive and summary statistics are made easier by the use of ICT. In this course you will learn how to use both spreadsheets and chart drawing programmes to help you with your statistical calculations and representations.

A large part of this course will be practically based. However, practical work does not have the same meaning in statistics as it does in science. In statistics it concerns handling real sets of data.

1.2 A practical question

A group of students have taken a spelling test. The average mark for the girls was 10 out of a possible 20. The average mark for the boys was 8 out of a possible 20. Does this mean that the girls, on the whole, did better in the test than the boys?

In Key Stage 3, you will have met the three definitions of average as

- the mode (the value which occurs most often)
- the median (the value which is halfway along the list when the numbers are written in order) and
- the mean (the total of the numbers divided by however many there are).

Suppose that 5 girls and 5 boys took the test and that their marks out of 20 were

     Girls: 9, 10, 10, 10, 11   Boys: 0, 0, 1, 19, 20

The students have the following averages

     Girls: Mode = 10   Median = 10   Mean = 10
     Boys: Mode = 0   Median = 1   Mean = 8

In every case, the average for the girls is greater than the average for the boys. Even the total mark for the girls (50) is more than the total mark for the boys (40), but we could hardly say that the girls did better than the boys when 2 of the boys (40% of them) scored very high marks. So you cannot infer that the girls did better in the spelling test than the boys.

Another piece of statistical information which should be considered in this simple case is the range, which you also met in Key Stage 3.

The range is a measure of the spread of the results. It is measured as

     range = highest result − lowest result

so, for the girls, range = 11 − 9 = 2
whilst for the boys, range = 20 − 0 = 20

Now we can describe the problem in more detail. A group of students has taken a spelling test. The mean mark for the girls was 10 out of 20. The mean mark for the boys was 8 out of 20. The range of marks for the girls was 2. The range of marks for the boys was 20. Discuss the possibilities for who did best in the test.

1.3 Statistical lines of enquiry

In many cases, statistics is about taking a line of enquiry and breaking it down into more manageable parts, perhaps even by asking sub-questions about it. For instance, the line of enquiry could be

     Who are, in general terms, the better at spelling, girls or boys?

from which you might test a suggestion, such as

     girls are better at spelling than boys,

for which you will gather some data, such as

     the results of a spelling test

which could be justified by

     comparing samples or types of sample.

You will take samples and obtain sets of results, make some form of comment, write a report, and justify your conclusions.

1.4 Cross-curricular uses of statistics

You may find that some of your other school subjects also make use of statistics. This is certainly the case in science, geography, economics, business studies and many other areas of study. It is possible that you may consider a hypothesis from another subject, collect data, and use it as a genuine cross-curricular topic by applying your knowledge of statistics in writing up your project.

2 The collection of data

In this chapter you will learn about the types of data, sampling data and how to collect data.

2.1 Types of data

Every day we are faced with collections of information. These may be in the form of observations, measurements or facts. We call this the collection of **data**.

To obtain data we must observe or measure something. This something is known as a **variable**. For example shoe size, height and eye colour are variables.

Variable	Measurement or observation
Shoe size	38, 39, 40, 41, 42
Height	169 cm, 178 cm, 183 cm, 185 cm
Eye colour	blue, brown, green

Shoe size and height are numerical measurements but the observations of eye colour are non-numerical. Shoe size and height are called **quantitative** variables. Eye colour is called a **qualitative** variable.

■ **Quantitative variables** have **numerical** observations or measurements.

■ **Qualitative variables** have **non-numerical** observations.

Consider the two examples of quantitative data: height and shoe size.

The height of people is measured on a continuous scale but shoe size can only take particular values.

■ **Continuous** data are measured on a scale and can take any value on that scale.

■ **Discrete** data are, in most cases, concerned with a number of countable values.

Gender and height are variables.

● Gender is *qualitative* and it is *discrete*.

● Height is *quantitative* and it is *continuous*.

We can sometimes regard continuous data as discrete. For example, a person's age can be regarded as discrete, if given to the nearest whole year, e.g. 16, 17.

Example 1

Mr Jones has just bought a new car.
Write down three variables, each associated with a car, which give
rise to each of the following types of data:

(a) quantitative, (b) qualitative,

(c) continuous, (d) discrete.

(a) mileage, number of passengers and engine size

(b) type, registration and colour

(c) age, length and miles per gallon

(d) number of seats, number of doors and purchase price

Exercise 2A

1   Billy spent last summer selling raffle tickets at a fairground.
    He sold red, green, blue and white tickets.
    (a) Explain whether the number of tickets he sold each day was
        discrete or continuous.
    (b) Explain if the variable 'number of tickets sold' is
        quantitative or qualitative.
    (c) Describe a qualitative variable associated with the tickets.

2   Penny spends 45 minutes writing an essay.
    Is the 45 minutes a discrete or a continuous variable? Explain
    your answer.

3   Joanne bought a coat last week.
    Describe a variable associated with the coat that is
    (a) discrete,
    (b) qualitative,
    (c) continuous.

4   Explain whether the total number of PCs bought this year is a
    discrete or a continuous variable.

5   Look at the picture of the London Eye.
    Write down the name of a variable connected to the Eye which
    gives data that is:
    (a) continuous, (b) discrete,
    (c) quantitative, (d) qualitative.

6   Karen says that her age is 16 years.
    Explain why this response is actually *continuous* data but may
    appear to be *discrete*.

In statistics, different variables form data that require different scales of measurement. We shall take a look at the main examples.

Categorical data

■ **Categorical data are data which have a scale grouped in different categories.**

Cars are categorised by insurance companies for insurance purposes. Here are some examples:

Group 1	Fiat Seicento	These are very small, fairly
Group 2	Ford Ka	cheap cars
Group 4	Honda Civic	Small family car
Group 11	Ford Mondeo	Medium sized family car
Group 13	Vauxhall Omega	Getting bigger …
Group 15	Jaguar XJS	… and quicker
Group 20	Ferrari 575	Very large or very fast!

Ranking data

Cars can be **ranked** or ordered in terms of cost when new, or engine size, or other features. For example, the car with the highest cost is ranked 1, the next highest cost is ranked 2, etc.

Interval scales

An **interval scale** is one where a single value or definition is used to represent all values between two end points. The age of racehorses is measured on an interval scale.
According to the rules all racehorses become officially a year older on 1 January each year. Horses are known as 1 year olds, 2 year olds etc. A race such as the Derby is for 3 year olds, but 3 year old horses could have real ages from 3 years to nearly 4 years old.

Ratio scales

A **ratio scale** (or logarithmic scale) is one in which equal distances show equal percentage changes. For example, one unit of length along the scale might represent 10, two units represent 100, three units 1000, etc.

The pH scale used in chemistry, as a measure of acidity or alkalinity, is a ratio scale.

Bivariate data

In many surveys undertaken in statistics you will examine how two variables are related or how one variable affects a second variable. Here are some examples:

- the price of a car and the age of the car,
- the ages of some students and their times to run a hundred metres,
- the weights and heights of a group of people.

Data of this type are usually called **bivariate**.

■ **Bivariate data consists of two related variables.**

Bivariate data can be discrete, continuous, grouped or ungrouped.

Very often in doing a survey or experiment or questionnaire you will need to **group** the data. There will be occasions when the data will be **ungrouped**.

Sometimes when you group data you lose some of the information. This will affect both the *accuracy* of any calculation and the *presentation* of the data.

Exercise 2B

1 How can a school register students using some form of categorical data?

2 Give examples of how each of the things below can be ranked to form a categorical ranking.
  (a) motorcycles   (b) houses
  (c) motor boats   (d) motor cars
  (e) students of mathematics (f) people in a club

3 Give two separate examples of interval data.

4 Give one example of bivariate data which will be:
  (a) discrete,   (b) continuous,
  (c) grouped,   (d) ungrouped,
  (e) discrete and grouped, (f) grouped and continuous.

5 Here are the ages, in years, of some people in a village.
  ```
  32  40  17  34  58  60  15  14  22  29
  44  18  26  31  36  42  18  23  25  38
  31  33  28  47  65  72  19  77  30  34
  37  34  58  56  60  63  42  15  82  17
  40  61  33  42  21  30  42  72  16  22
  ```

  Group these ages according to the categories:
  0 to 9, 10 to 19, 20 to 29, 30 to 39, 40 to 49, 50 to 59, 60 to 69, 70 to 79, and 80 to 89.

2.2 Collecting data

Before you collect any data, you need to have an aim. If you do not have an aim, or the aim is not clear, the data will be meaningless.

Primary or secondary data?

- Data that is collected by or for the person who is going to use it is called **primary data**.

If you collect the birth dates of your class by asking each student, this would be primary data.

- Data that is not collected by or for the person who will use it is called **secondary data**.

If you collect the birth dates by looking at the school records, this would be secondary data.

Both forms of data have advantages and disadvantages.

Data	Advantages	Disadvantages
Primary data	You know how they were obtained. Accuracy is known	Time consuming. Expensive.
Secondary data	Easy to obtain. Cheap to obtain. The data might be out of date. May have mistakes.	Method of collection is unknown.

Experiments and surveys

■ **Data can be collected using an experiment or a survey.**

In a **statistical experiment**, one of the variables will be controlled while its effect on the other variable is observed.

■ **The controlled variable is called the explanatory or independent variable.**

■ **The effect being observed is called the response or dependent variable.**

A **survey** can be a particularly useful mechanism for collecting data, in particular data which are likely to be personal. A survey may fail, because you can never guarantee that people are being truthful.

Here are some of the main survey methods:
- postal surveys – in these a survey is sent to an address
- personal interviews – in which people are asked questions; this is often used in market research and similar activities

- telephone survey – in which the telephone is used; it is often a special type of personal interview
- observation – this usually involves the monitoring of information or behaviour.

Each method has advantages and disadvantages.
The costs of conducting a survey by each method will vary greatly.
The person conducting the survey will need to consider a number of these factors before deciding which method to choose.

As with most of statistics, you will never reach a perfect conclusion; but it is possible to get some form of overall view.

This table very briefly sums up the main advantages and disadvantages of each type of survey.

Method	Advantages	Disadvantages
Postal	Large amount of data. Relatively cheap to do.	Possible poor response. Limited in the data that can be collected.
Personal interview and telephone survey	Good response rate. Can ask lots of questions.	Expensive to do. Interviewer can influence the responses given.
Observation	Very systematic and mechanical.	Results can be dependent on chance.

Example 2

James is carrying out an experiment to see which class in Years 10 and 11 at his school is best at arriving on time for various lessons. What explanatory (independent) and response (dependent) variables could James use?

The explanatory variable will be the classes in years 10 and 11.
The times of arrival at each lesson will be the response variable.

Example 3

A creative design company is given a contract to design and market a new lifestyle magazine aimed at younger women.

(a) Explain how and why they could use both primary and secondary data.
(b) Explain a possible method of collecting primary data.

(a) Primary – survey to find out which features of a magazine are most and least attractive.
Secondary – refer to other publications to see what female readers like to read about in magazines.

(b) By post, personal interview or by telephoning some potential readers.

Exercise 2C

1 The police are making a study of crime rates in a town. They have a number of officers wearing plain clothes who study the possible crimes that take place on market day.
Explain why these officers will be likely to provide the police with primary data.

2 How could you use secondary data to help design a new hot-water bottle?

3 Damion has a theory that taller people will, in general, be better sprinters than shorter people. He tests his theory by conducting an experiment.
What are the explanatory and response variables he should measure?

4 A sweet company wishes to produce a new version of fruit gums. They can use primary data, secondary data or a mixture of both.
You have to advise the company about the data they use. Write a few notes advising the chairman of the company about the sort of data they should consider.

5 A company is considering building a new swimming pool in the town centre.
Suggest a method of collecting primary data to provide information to help the company decide whether to build the swimming pool or not. Explain your answers.

6 James and Colin wish to predict the winners of the next World Cup in football.
James looks at the World Cup results from 2002, which had Brazil as the winners.
Colin looks at the table of World Cup results from when the competition started in the nineteen thirties.
(a) What types of data are they considering?
(b) Whose opinion would you trust most and why?

7 Joanne wishes to carry out a survey to see how much money people spend on food each week.
Advise Joanne on the relative merits of conducting her survey
(a) by interview,   **(b)** by telephone.

8 You are undertaking a survey to see how much money people will spend buying a car. Give one reason why you would choose to conduct a personal interview rather than a postal survey. Also explain why you might not consider conducting personal interviews.

2.3 Population and sampling

When you undertake a survey the first thing to do is to correctly identify the **population**.

■ **A population is everything or everybody that could possibly be involved in an investigation.**

The definition of 'population' can vary.

The owner of Highfield Motor Company wishes to work out the average miles travelled by the cars in the company's garage. The population being considered is all the cars in the garage.

If you are going to ask questions of some of the people in this picture, the population being considered is all the people at the bus stop.

Census data

■ **Census data contains information about every member or element in a population.**

Workers for central and local government, health authorities, banks, the police and other organisations often use census information to help allocate resources or plan services for all people under all circumstances.

Almost certainly the best known census, is the Government's National Census. This takes place every ten years. All people in the country are identified and asked certain questions.

A census is usually only practical when the population is small and well known. For example, in a small company, the number of hours of overtime worked by the employees could be recorded.

A census is not always possible. For example, if you were testing a component to destruction. An example of this is testing climbing ropes to find out at what load they break.

Sample data

One of the most often used statistical methods is the taking of a **sample**. This is usually done when the population is large and collecting data might cost too much or might take too long.

■ **When sample data are collected, information is taken from part of the population. The information is then used to make conclusions about the whole population.**

Always remember that sampling can provide incorrect results or conclusions. Nothing can ever be *proved* by sampling, but it can provide you with the most likely alternative or a set of likely alternatives. You can, in most cases, attach probabilities to the set of alternatives.

Probabilities are covered in more detail in Chapter 8.

At all stages of taking a sample you need to be sure of what you are doing and the sample **size** being taken. Sample size can vary, but usually the larger it is the more reliable the results will be. It is important, however, to remember what the sample is being used for and to balance the size of the sample against the accuracy you need. This is a major feature of using statistics in any situation.

A census collects information from every member of a population. A sample collects information from only part of a population.

You should always try to ensure that any sample you take is free from **bias**. This simply means that your results do not have any sort of distortion.

Bias in sampling may result from these factors:

● Failing to correctly identify the population.
  There is little to be gained from a survey about what and how women think if you ask men.

● Choosing a sample which is not representative.
  For example, in a survey about the reliability of all cars, only including cars which are less than three years of age.

● Failure to respond to a survey.
  Many people do not fill in responses to questionnaires sent through the post.

● Asking ambiguous or misleading questions.
  Try not to ask questions such as, 'Are you tall?'.

● Careless answers to questions.
  People may make honest errors in answering questions.

● Dishonest answers to questions.
  There are people who think it is clever to provide dishonest answers to certain questions, such as 'How old are you?'.

- Errors in recording answers.
  It is easy to make mistakes in recording answers, such as writing 25 when you mean 52.
- The person being asked having a personal interest.
  For instance, asking a manufacturer of alcopops which brand they think is most popular.
- External factors influencing the data being collected.
  You may have no control over factors such as the temperature or the weather, which may affect opinions on the topic that you are surveying.

The advantages and disadvantages of using a sampling technique and a census are summed up in the table below.

Method	Advantages	Disadvantages
Census	Unbiased. Accurate. Takes into account the whole population.	Time consuming. Expensive. Difficult to ensure that the whole population is surveyed.
Sample	Cheaper. Less time consuming. Less data to be considered.	Not completely representative. Possibly biased.

Exercise 2D

1 A new canteen is about to open at Edwell Secondary School.
   (a) Give a reason why the canteen managers should conduct a census survey to determine what the canteen should consider offering on its menu.
   (b) Identify the population for this survey.

2 Electrobatts, battery producers, conduct a statistical experiment to find out how long a new battery will last.
   (a) Give a reason why a census should not be carried out.
   (b) Identify a possible population to use in this experiment.

3 Joanne is conducting a survey into how women travel to work. What are the possible errors Joanne could make in her sampling technique? Give three examples.

4 George is doing a project about house prices in various parts of England.
   Give an example why he:
   (a) should not conduct a census of all the house prices in England,
   (b) might take a census of all the house prices where he lives.

5   (a) State a survey that you might wish to carry out.

   (b) Give one question that you might ask, knowing that it could lead to:

      (i) a biased response,

     (ii) a dishonest answer,

   (iii) an answer mainly from one gender.

2.4 Randomness

Whenever you need to take a sample it is likely that the sample will need to be a **random** sample.

■ **In a random sample every item in the population will have an equal chance of being selected.**

We shall take a look at the concept by considering the problem of taking a random sample of advertisements for motorcycles. Suppose that you have 1000 advertisements for the motorcycles. A sample of 20 is required.

Method 1

● List the motorcycles from 1 to 1000.

● Now choose a table of random numbers. For example:

   943061   231546   627349   554710   119537   027184 and so on.

● Split these into three-digit numbers:

   943 061 231 546 627 349 554 710 119 537 027...

● To choose your sample of 20, select from the list of 1000 motorcycles those listed as numbers 943, 061, 231, etc., until you have a list of 20.

> If you were choosing from a population of 50, then you would split into two-digit numbers and the choices would be 94, 30, 61, 23. You would ignore 94, 61 and all others greater than 50. You would also ignore any number that is repeated.

Method 2

● List all the motorcycles from 1 to 1000.

● Put the numbers from 1 to 1000 in a hat.

● Shake the hat so that the order of the numbers cannot be determined.

● Withdraw 20 numbers from the hat.

Method 3

● List all the motorcycles from 1 to 1000.

● Obtain a calculator.

● Input 1000 and press SHIFT then press Ran# and now press =

● Read off the number (say, 214).

> This may be different on your calculator.

Exercise 2E

1   Tim is asked to take a random sample of 25 students from the registration roll at his school.
    He attempts to do so by:
    - listing all their names in order,
    - rolling a die,
    - selecting the student shown by the number on the die (i.e. if the die shows 4 he selects the student numbered fourth in the list),
    - rolling the die again. If it shows 3 he selects the 4th + 3rd = 7th name on the list) and so on.

    (a) Give two reasons why Tim's method will **not** give him a random sample.

    (b) Write down a correct method which could be used to take a random sample of the students at Tim's school.

2   Alice needs to select a random sample of 40 students from her school of 1200 students. She decides to use her calculator to do this.
    Explain how she can do so.

3   Below is an extract from a table of random numbers.

    335217   045178   627341   532715   823859   482082   342173
    451739   936415   526338   127642   137284   463919   394821
    264519   143857   012653   628491   558317   316832   229103

    Use this table of random numbers to make a selection of ten random numbers each less than 50.

4   Sadiq wishes to take a random sample of 20 cars.
    He has five magazines from which he can extract the data. These are:

Magazine	Number of pages of adverts	Number of cars per page advertised
A	100	20
B	50	10
C	200	20
D	100	30
E	70	15

    Explain three methods which will guarantee that Sadiq successfully chooses a random sample of cars from those advertised.

2.5 Methods of sampling

We usually take a **random sample** from a population to ensure that every item in the population has an equal probability of being selected in the sample. If this does not happen then the sample will be biased and any results drawn from the sample will be unrealistic.

In most cases, the larger the sample size, the more representative the statistics will be of the whole population. This is not always the case, but it is a fairly good premise from which to start.

■ **A random sample is a sample which is chosen without any conscious decision about which items from within the population are selected.**

The identification of the population is determined by a **sampling frame**, which simply consists of all the items in the population. Using the sampling frame ensures that every item in the population has a chance of being sampled.

It is important that you should be able to identify a sampling frame. This is important especially in the coursework.

Simple random sampling

A sample size of *n* is called a **simple random sample** if every possible sample of size *n* has an equal chance of being selected. In practice, this is achieved by giving each member of the population an equal chance of being selected. There are many ways of doing this. You could give each item a number and then select the numbers of the items to be included in the sample by:

- using a random number table,
- using a random number generator on a scientific calculator,
- using a computer to choose the numbers,
- putting the numbers in a hat and then selecting however many you need for your sample.

Any of these methods is best suited to a relatively small population where the sampling frame is well determined.

Stratified sampling

To do this, first divide the population into **strata** or categories by gender (male or female), age, social class, earnings, etc. Then from within each stratum, take a random sample. The size of each sample is in proportion to the relative size of the stratum from which it is taken.

This method ensures that a fair proportion of responses from each group is sampled.

Example 4

The headteacher of a school of 1000 students wishes to take a sample of 60 students.
The number of students in each year is as follows:

	Year 7	Year 8	Year 9	Year 10	Year 11
Students	250	250	200	150	150

How many students should be included in the sample from each year?

The number of students is $250 + 250 + 200 + 150 + 150 = 1000$
So the head's sample will be, for Year 7:

$$\frac{250}{1000} \times 60 = 15$$

For all years, the numbers in the samples will be:

Year 7	Year 8	Year 9	Year 10	Year 11	Total
15	15	12	9	9	60

The headteacher might wish to split the sample into boys and girls from each year group. So the sampling could be:

	Year 7	Year 8	Year 9	Year 10	Year 11
Girls	8	7	6	5	4
Boys	7	8	6	4	5

This is an example of a stratified sample with *more than one* category being used.

Systematic sampling

A sample that is obtained by choosing at regular intervals from an unordered list is called a **systematic sample**. For example, if you wish to get a sample of 20 students from a year group register of 100 students, you could take every 5th student from the register. The first student to be picked is a random sample between 1 and 5. So if 3 is picked, you would take the 3rd, 8th, 13th, 18th, 23rd etc. from the list. Systematic sampling is used when a population is very large. It is simple to use, but it may not always be truly representative.

$\frac{100}{20} = 5$

Cluster sampling

The population being sampled is split into groups or clusters. The clusters to be sampled are randomly chosen and every item in the cluster is looked at.

You should ensure, unless you have a good reason to do otherwise, that the clusters are of more or less equal size. It is best if a large number of small clusters is formed since this minimises the chances of the sample being unrepresentative.

The method is very popular with scientists.

Quota sampling

This method of sampling is most commonly used in market research.

Instructions are given about the **quota** or amount of each section of the population with a certain characteristic, such as age, that is to be sampled.

No sampling frame will be required, which is an advantage; but there is a major disadvantage in that the choice of the sample is the sample surveyor's decision which can lead to questions of bias.

Convenience sampling

In this the most convenient sample is chosen. For, say, a sample of 50 students in a school, the sample may be simply the first 50 names on the register.

The method is very quick and does not need too much time to organise. It does have a major disadvantage in that it can lead to immeasureable levels of bias and, of course, the sample is likely to be unrepresentative.

Combining methods

Sometimes more than one method can be used. For example, opinion polls used in politics use a combination of quota and cluster sampling to construct a large-scale poll.

The criteria for the selection of a sample in any national opinion poll include gender, age, geographical area, social background, economic background and others.

While a sample size may well be large, it is often based on a very small proportion of the population.

One of the major disadvantages of the conclusions drawn from any opinion poll is that opinions can change, sometimes almost immediately, over a period of time.

Each method of sampling has a particular drawback. Sampling methods can create bias, influences, etc., which is inconvenient. You will need to be aware of these factors. You need to learn how to avoid them and how to set up projects or lines of enquiry that minimise the effect of these factors.

> If there are n items in the population an appropriate sample size will be $\sqrt{n}$.

Exercise 2F

1   Yvonne is doing a survey to find out how far people travel each day to get to school. She asked every person at the school bus stop one day.
    (a) Identify the sampling frame she used.
    (b) Explain, giving your reasons, why the sample is biased.
    (c) Explain a better method she could use to choose the sample.

2   Jack wishes to find out how much people in Britain are prepared to spend on a weekend break. He asked people in his village.
    (a) Identify Jack's population.
    (b) Explain why Jack's sample is likely to be biased.
    (c) Suggest a better way of obtaining the information and explain why it is better.

3   Give an example of a:
    (a) simple random sample,
    (b) systematic sample,
    (c) cluster sample,
    (d) quota sample,
    (e) convenience sample.

4   John is doing a statistical project. He decides to take a sample of 20 people.
    The sample he chooses is the first 20 names on a register.
    (a) What sort of sample is he choosing?
    (b) What are the two major problems with this form of taking a sample?
    (c) Suggest a better method of choosing a sample.

5   Jenny is investigating the statement, 'Children do not enjoy sport'.
    She does so by surveying the attitudes to sport amongst the pupils at her school.
    (a) Comment on the method.
    (b) Give one alternative method of surveying opinions.

6   Mr Hardy wishes to use a stratified sample of 25% of the pupils at Shimpwell First School which has a total of 160 pupils.
    (a) How many pupils will be in the sample?

    Shimpwell First School has the following numbers of pupils:

Reception Year	Year 1	Year 2
40 pupils	60 pupils	60 pupils

    (b) Work out the numbers of pupils of each age group in the stratified sample.

7   (a) Describe how you could take a systematic sample of 50 cars that you see driving along your road.

    The sample is part of a project looking at which cars people drive.

    (b) Explain carefully the merits and disadvantages of this sampling method.

8 Explain three methods Jennifer could use to estimate the number of blades of grass on her lawn. State clearly the reasons for each method of estimating and give the reasons for the choice.

9 The headteacher of Long Melling Secondary School wishes to interview a stratified sample of 50 pupils at the school.
The roll of the school is

	Year 7	Year 8	Year 9	Year 10	Year 11
Girls	125	130	100	100	75
Boys	125	120	100	100	75

Work out the numbers of pupils of each gender in each year to be in the sample. You will need to round some decimals either up or down or make alternative arrangements.

2.6 Questionnaires

Writing **questionnaires** is an important part of statistics. A questionnaire is used as a method of collecting data. Sometimes the questions can be factual, such as, 'How old are you?' or they might in other cases ask for an opinion, such as, 'What are your favourite colours?'.

■ **A questionnaire is a set of questions designed to obtain data from a population.**

Anyone who answers a questionnaire is called a **respondent**.

All questionnaires should:
- be clear as to who should complete them,
- be as brief as possible,
- start with simple questions to encourage the person who is completing it,
- be clear *where* the answers should be recorded,
- be clear *how* the answers should be recorded,
- be able to be answered quickly.

These objectives are likely to be achieved by:
- asking short questions,
- using simple, clear language,
- being free from bias,
- being unambiguous,
- not asking leading questions,

and, when you are doing a survey, the questionnaire should be
- useful and relevant to the survey you are undertaking.

When you design a questionnaire there are two types of questions you should use.

You could ask an **open** question or a **closed** question.

■ **An open question is one that has no suggested answers and will give people a chance to reply as they wish.**

An open question could reveal an answer or response, or set of answers or responses not considered by the person asking the question. It could, in certain circumstances, lead to a whole new line of enquiry or even a new statistical project.

■ **A closed question has a set of answers for the respondent to choose from.**

A closed question has the advantage that it is easier to summarise the data being collected and to make possible comparisons when needed.

There will be times when a questionnaire makes use of **opinion scales**. An opinion scale asks people to state their opinion about something. Most people will give an opinion near to the middle because they do not wish to appear extreme.

Here is a typical question and an opinion scale:
'England is a nation of dog lovers'.

☐	☐	☐	☐	☐
Strongly agree	Agree	No view	Disagree	Strongly disagree

One of the main problems with opinion scales is that most people will answer somewhere near the middle: they are unlikely to have a very strong opinion either way. If there are boxes, having an even number can help a little, because then nobody can go for the middle box.

In some cases the respondents may be asked to tick a box. This should be simple, as this question shows.

Tick your age in the boxes below:

☐	☐	☐	☐
Under 20	21 to 30	40 to 50	50 to 60

The boxes could be
☐   ☐   etc.
20 and   21–30
under

However, in this case,

• someone aged 20, or between 30 and 40, cannot tick a response,
• someone aged 50 could tick two boxes.

When you design a questionnaire you need to make sure that the answers are clear.

In an actual survey you may decide to use closed questions. If you do, then you must ensure that the choice of answers reflects the likely responses of those being questioned.

Closed questions have clear cut answers.

Whenever you undertake a survey or experiment it is sensible to do a **pilot survey**. A pilot survey is one that is carried out on a very small scale to make sure the design and methods of the survey are likely to produce the information required. It should identify any problems with the wording of questions, likely responses etc.

■ **A pilot survey is a small scale replica of the survey or experiment that is to be carried out.**

Very few people will be willing to provide information about themselves or their feelings. Many people resent being asked questions which are personal or sensitive.

Such questions should be avoided if possible, but in certain cases they cannot. If it is necessary to ask highly personal questions then you should leave them until the end of a questionnaire.

At times you can use the **random response** technique to ensure that you obtain truthful answers to very sensitive questions. Suppose you wanted to find out if any students in Year 7 smoked. Ask them:

'Toss a coin. If it shows heads then tick the Yes box below. If it lands tails then please answer the question:

Have you ever smoked a cigarette or cigar or a pipe? Yes ☐ No ☐

If you ask 100 Year 7 students then approximately 50 will answer yes. But if you find that there are 55 ticks in the yes box then it means that about 5 (or 55–50) Year 7 students have smoked. So you could say that as far as the students in Year 7 are concerned, about 1 in 10 (or 5 in 50) have tried smoking at some time or other.

Exercise 2G

1   Kerry is writing a questionnaire about people's ages.
In it she asks the question:
How old are you?   Young ☐   Middle-aged ☐   Old ☐
   (a) What is wrong with the question and answers?
   (b) Rewrite the question and answers.

2   Steve is doing a survey on football teams.
He writes a question and answer box:
How often do you watch a football match?
Never ☐   Once a week ☐   When I can ☐
   (a) Give your reasons why Steve should not ask the question in such a way.
   (b) Write a better question for Steve to use.

3   George is doing a survey into people's television viewing habits.
He wants to know how much time people spend watching television.
Write down three questions that George should ask in his survey.

4   Angela is doing a questionnaire because she wants some information about the weights of people in her school.
She asks them:

How much do you weigh?

(a) Explain why that is not a good question to ask.

(b) Write a good question and give a set of response boxes with it.

5   Write a questionnaire to find out the favourite cans of drink bought by the students at your school.
State clearly the sampling frame, the sample size and how you would obtain the sample.
Describe how you could use a pilot survey.

6   A market research company intended to ask the following question in a questionnaire:

How old are you?

(a) Say why this question is unacceptable.

(b) Rewrite the question in such a way that it will be acceptable.

2.7 Experimental design

Here are the five methods you should consider whenever you undertake an experiment as part of a statistical investigation.

Before and after experiment

You could conduct a before and after experiment to help you make a judgement on the influences that a factor could have.

A typical example could be looking at the effects of the 2003 war in Iraq.

Capture–recapture method

The capture–recapture method is used to estimate the size of a self-contained population. A sample is taken from the population and tagged before being replaced in the population. Later, a second sample is taken and the number of tagged members recorded. From this you can estimate the total size of the population.

For example, to estimate n, the number of birds living near your home, you could capture 20 birds. Tag or ring the birds and release them. Later, capture 100 birds. If you find that, say, 10 of these are tagged, you can estimate n from

$$\frac{10}{100} = \frac{20}{n}$$
$$n = 200$$

Control group

A control group is often used when we want to test the effect of various factors in an experiment. The control group is randomly selected, and is not subjected to any of the factors you wish to test.

For example, you might use this method to test the effect of a new drug. The control group of patients is given an inactive substance instead of the drug.

Data logging

This is a mechanical or electronic method of collecting primary data. You programme an instrument to take readings at set intervals.

For example, data logging can be used to measure pulse rates after exercising.

Matched pairs

Here you use two groups to investigate the effect of a particular factor. You need to ensure that both groups to be tested have everything in common except the factor being tested or studied. Identical twins can be very important in these sorts of experiments.

You may wish to replicate, or repeat, some observations. This is called **replication**.

You may wish to ensure that observations are selected at random. This is called **randomisation**.

Exercise 2H

1   For each of these six experiments decide which experimental method you would use.
   (a) An estimate of the number of sparrows in a rural area is to be made.
   (b) The heart rate of people of various ages is to be monitored when they run a mile.
   (c) The effect on the learning of children of the long summer break from school is to be tested.
   (d) An assessment is to be made on the effects of watching TV on students' learning.
   (e) A brand new treatment for rabies is to be tested.
   (f) An investigation of the effect of the environment on personality is to be made by studying pairs of identical twins reared apart from an early age.

2   Sixty birds at a bird sanctuary are caught. They are tagged and returned to the sanctuary. A second sample of 100 birds is caught later. Of these 100 birds, 20 have a tag.
   Estimate the number of birds in the sanctuary.

3   Write down one possible method of using
   (a) a before and after experiment,
   (b) a capture–recapture method,
   (c) a control group,
   (d) data logging,
   (e) matched pairs.

   (Do not give an example already covered in the text.)

4 Which method would you use to test the effect of drinking alcohol on driving?

5 The population of fish in a lake is to be estimated.
Explain how you could do it. How could you be sure that no fish die as a result of being sampled?

2.8 Accuracy

In statistics you collect data. On many occasions these data will be for something which is continuous. Whenever data are continuous they will be measured and rounded to the nearest sensible unit.

■ **A measurement given correct to the nearest whole unit can be inaccurate by up to $\pm\frac{1}{2}$ unit.**

■ **The upper bound is the largest value that a measurement could be. The lower bound is the smallest value that a measurement could be.**

Age is perhaps the main exception to this. If someone truthfully tells you that their age is 15 then they could be anything from 15 years and 0 days to 15 years and 364 days old. We would not describe someone of 15 years and 11 months as being 16 years old, although we might say that they are *nearly* 16 years old.

When we say that a measurement $x = 5$, where x is continuous, is correct to the nearest whole unit then the maximum possible error in x is $\pm\frac{1}{2}$ unit.

$$x = 5 \pm \tfrac{1}{2} \quad \text{or} \quad 4\tfrac{1}{2} \leqslant x < 5\tfrac{1}{2}$$

So the lower bound for x is $4\frac{1}{2}$ and the upper bound is $5\frac{1}{2}$ units.

Significant figures

If the length of the page of a book is given as 26.4 cm it is given to 3 significant figures. The actual length could be $\geqslant 26.35$ but < 26.45. The length is 26.4 cm correct to the nearest 0.1 cm.

Example 5

The distance John travels to school is $4\frac{1}{2}$ miles to the nearest $\frac{1}{2}$ mile. Work out the lower and higher bounds for the distance John travels.

The maximum possible error will be $\pm\frac{1}{4}$ mile.
Higher bound $= 4\frac{1}{2} + \frac{1}{4} = 4\frac{3}{4}$ miles
Lower bound $= 4\frac{1}{2} - \frac{1}{4} = 4\frac{1}{4}$ miles

Example 6

A rectangular room is measured as 12 metres by 10 metres. Both measurements are given to the nearest metre.

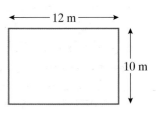

Work out:

(a) the upper bound for the area of the room,

(b) the lower bound for the area of the room.

The area of the room is length $\times$ breadth.

The upper bound of the length of 12 metres is 12.5 metres.
The upper bound of the width of 10 metres is 10.5 metres.

So the upper bound for the area will be $12.5 \times 10.5 = 131.25$ square metres.

The lower bound of the length of 12 metres is 11.5 metres.
The lower bound of the width of 10 metres is 9.5 metres.

So the lower bound for the area will be $11.5 \times 9.5 = 109.25$ square metres.

Exercise 2I

1 Find the upper and lower bounds for:
   (a) a TV which costs £200, correct to the nearest £5,
   (b) a car which costs £10 000, correct to the nearest £500,
   (c) the height of a wall labelled as 4 metres, correct to the nearest metre,
   (d) a garden tree whose height is given as 5 metres, correct to the nearest metre,
   (e) a railway journey of 125 miles, correct to the nearest mile,
   (f) a boy's height of 175 centimetres, correct to the nearest centimetre,
   (g) the age of a man of 63 years,
   (h) the distance from London to Manchester, which is given as 200 miles, correct to the nearest 5 miles.

2 Work out the upper and lower bounds for:
   (a) the area of a rectangular carpet which measures 5 metres by 4 metres (both measurements are correct to the nearest $\frac{1}{2}$ metre),
   (b) the weight of a mug of coffee which is given as 100 g correct to the nearest 5 g,
   (c) the volume of a cube which is quoted as having a side of length 6 cm correct to the nearest centimetre,
   (d) Tom's age of 15 years 6 months and 24 days,
   (e) the volume of a cylinder with base of radius 4 cm and height 10 cm; the radius of the base and the height are both correct to the nearest centimetre.

3 The two variables x and y are given as

$x = 5$ and $y = 3$

Both can be subject to an error of 0.5 units.

Work out:

(a) the upper and lower bounds for x,

(b) the upper and lower bounds for y,

(c) the upper and lower bounds for

$$\frac{x + y}{x}$$

4 $s = 5$, correct to 1 significant figure.
$t = 0.3$, correct to 1 significant figure.
Calculate **(a)** the upper bound and **(b)** the lower bound of

$$t - \frac{s - 6}{s}$$

2.9 The presentation of data

Raw data are usually collected by either counting or measuring. It is sensible to collect them by using a frequency table which is easy to read and also to analyse. You will learn about frequency tables in Chapter 3.

Revision exercise 2

1 There are three horses in a field.

discrete   quantitative   qualitative   continuous   cumulative

Use one of the words above to complete these sentences:

(a) The colour of the horses is .... data.

(b) The number of the horses is .... data.

2 The head teacher in a primary school took a random sample of 10 boys and 12 girls from all the children in the school.

(a) Write down the sampling frame used by the head teacher.

(b) Write down the type of sampling scheme used.

The head teacher asked each of these 22 children the following question as part of a questionnaire.

'You go to bed before 9.00 p.m., don't you?'

(c) Give one reason why the head teacher should not have asked the question in this way.

3 Give an example to illustrate the use of each of the following scales of measurement:

(a) categorical,   **(b)** rank,   **(c)** interval.

4 The table shows the number of students in each year and in the sixth form at Wellup High School.

	Year 7	Year 8	Year 9	Year 10	Year 11	Sixth Form	Total
Girls	125	104	136	99	104	184	752
Boys	127	116	133	89	102	178	745
Total	252	220	269	188	206	362	1497

George wants to find out the students' opinion of the school's plan to change the school day.
He decides to select a sample of 150 students and ask them a series of questions.

(a) Explain why 150 students will be a suitable sample size.

(b) Work out the number of girls and the number of boys he should select from Year 9.

5 The table shows the number of students in each of the four Year 11 maths classes in a school.

A sample of size 30 is to be taken from Year 11. Omar suggests that 3 of the classes are chosen at random and 10 students selected at random from each class.

Maths class	Number of pupils
Class 1	35
Class 2	25
Class 3	20
Class 4	10

(a) Would this method give a random sample? Explain your answer.

Nesta suggests a stratified sample of size 36 from the whole Year 11 using classes as the strata.

(b) How many students from Class 1 should be in the sample?

6 Here is part of the roll for Mayfield High School. It shows the numbers of boys and girls in Years 10 and 11.

Vijay wishes to take a representative sample of 60 students from these two year groups.

Explain how this can be done.

	Year 10	Year 11
Girls	106	84
Boys	94	86
Total	200	170

7 Mr Smith is going to sell drinks on his coaches.
He wants to know what type of drinks people like.
He wishes to use a questionnaire.

Design a suitable questionnaire, which he could use to find out what type of drink people like.

8 There are 800 pupils at Hightier School. The table shows information about the pupils.

An inspector is carrying out a survey into pupils' views about the school.
She takes a sample, stratified both by year group and by gender, of 50 of the 800 pupils.

Year group	Number of boys	Number of girls
7	110	87
8	98	85
9	76	74
10	73	77
11	65	55

(a) Calculate the number of Year 9 boys to be sampled.

Toni stated 'There will be twice as many Year 7 boys as Year 11 girls to be sampled.'

(b) Is Toni's statement correct?
You must show how you reached your decision.

Summary of key points

1 Quantitative variables have numerical observations or measurements. Qualitative variables have non-numerical observations.

2 Continuous data are measured on a scale and can take any value on that scale.
Discrete data are, in most cases, concerned with a number of countable values.

3 Categorical data are data which have a scale grouped in different categories.

4 Bivariate data consists of two related variables.

5 Data can be collected using an experiment or a survey.

6 The controlled variable is called the explanatory or independent variable. The effect being observed is called the response or dependent variable.

7 A population is everything or everybody that could possibly be involved in an investigation.

8 Census data contains information about every member or element in a population.

9 When sample data are collected, information is taken from part of the population. The information is then used to make conclusions about the whole population.

10 In a random sample every item in the population will have an equal chance of being selected.

11 A random sample is a sample which is chosen without any conscious decision about which items from within the population are selected.

12 A questionnaire is a set of questions designed to obtain data from a population.

13 An open question is one that has no suggested answers and will give people a chance to reply as they wish.

14 A closed question has a set of answers for the respondent to choose from.

15 A pilot survey is a small scale replica of the actual survey or experiment that is to be carried out.

16 A measurement given correct to the nearest whole unit can be inaccurate by up to $\pm\frac{1}{2}$ unit.

17 The upper bound is the largest value that a measurement could be. The lower bound is the smallest value that a measurement could be.

3 Representing and processing discrete data

3.1 Tally charts and frequency tables

Before data have been processed they are known as **raw data**. It is not easy to see patterns or trends in raw data. To make the data easier to interpret they can be sorted into a **tally chart** or **frequency table**.

Example 1

Here are the ages of 40 children who attend a youth club.

12 14 14 15 13 13 14 11 14 13
14 13 12 13 14 12 15 16 13 14
15 13 14 12 14 13 11 14 13 14
15 13 12 11 14 15 14 13 14 12

Sort the data into a frequency table.

Ages can be treated as discrete or continuous data. In this chapter they will be treated as discrete.

Age	Tally	Frequency
11					3
12	ℍℋ I	6
13	ℍℋ ℍℋ I	11
14	ℍℋ ℍℋ ||||	14
15	ℍℋ	5
16			1
Total		40

Tally marks are made in groups of five, with the fifth tally mark drawn through the other four. This makes the tallies easier to count.

Add up the tally marks to find the frequency.

The first column shows us that the ages range from 11 to 16.

In the second column each child is represented by one **tally mark**.

The **frequency** column tells you how often each age occurred.

■ **A tally chart, or frequency table, can be used to process raw data, making it easier to spot patterns.**

Exercise 3A

1   Here are the colours of 50 cars in a car park.

red	white	white	blue	white	yellow	white
blue	white	red	blue	red	yellow	white
blue	blue	red	black	white	white	black
white	white	blue	red	white	white	black
yellow	blue	white	red	white	red	red
red	red	black	red	blue	black	blue
blue	red	black	white	white	black	white
white						

Copy and complete the frequency table to show the colours of the cars.

Colour	Tally	Frequency
Black		
Blue		
Red		
White		
Yellow		
Total		

2   Last season Woodbank football team played 30 matches.
Here is a list of the number of goals they scored in each match.

```
2  3  0  1  2  4  3  0  2  1
2  2  1  3  0  2  4  0  2  1
0  5  3  2  1  4  1  2  0  1
```

Design and complete a frequency table to show the distribution of the number of goals scored.

3   A company sells drawing pins in boxes. They claim that there are 50 drawing pins in each box.

Des checks the contents of 20 boxes. The number of drawing pins in each box is shown below.

```
48  52  51  50  50  50  49  50  50  51
53  50  51  52  49  50  50  52  49  50
```

(a) Design and complete a frequency table to show the number of drawing pins in each box.

(b) Do you think that the company's claim that 'there are 50 drawing pins in each box' is reasonable?

4   The length of a word can be measured by counting the number of letters it contains. Count the length of each of the 52 words in the yellow boxes on page 29. Record your answers in a frequency table.

5 For this question, you will need three coins.
Toss all three coins at the same time, and count how many coins
land on 'heads'.
Conduct this experiment 25 times.
Record your answers in a frequency table.

3.2 Grouping data

Sometimes the data have a wide range of values, with few values
that are the same. Data can be sorted into groups or **classes**. This
shows the distribution of the data more clearly, making it easier to
spot patterns.

Example 2

Sort the following data into groups.

0	7	34	40	52	53	52	24	48	32
56	3	2	35	42	55	56	14	23	34
51	54	6	6	49	48	55	57	12	29
62	63	73	4	22	45	44	45	47	5
71	79	70	77	3	28	44	47	49	2
72	78	71	72	75	9	27	43	46	9

A frequency table for these data is shown below.

Age	Tally	Frequency
0–9	𝖧𝖧 𝖧𝖧 II	12
10–19	II	2
20–29	𝖧𝖧 I	6
30–39	IIII	4
40–49	𝖧𝖧 𝖧𝖧 IIII	14
50–59	𝖧𝖧 𝖧𝖧	10
60–69	II	2
70–79	𝖧𝖧 𝖧𝖧	10
Total		60

Each class in the table in Example 2 covers ten different integers.
The intervals 0–9, 10–19, etc. are called **class intervals**.

■ **When the data are widely spread, you should group the data
into classes.**

When grouping data, you must give careful thought to the number
of class intervals you use, and the width of the intervals.
The next frequency table shows the same data, using wider class
intervals.

Age	Tally	Frequency
0–19	卌 卌 IIII	14
20–39	卌 卌	10
40–59	卌 卌 卌 卌 IIII	24
60–79	卌 卌 II	12
Total		60

If you do not use enough class intervals, or the intervals are too wide, you can lose some of the important details. The original table showed that many of the data were below 10. This information was lost in the second table.

Similarly, if too many intervals are used, each frequency will be too small. This could hide any patterns or trends.

■ **Choose an appropriate number and width of class intervals in order to make any patterns and trends more obvious. Between 5 and 10 class intervals will often be appropriate.**

Exercise 3B

1 A mathematics test is marked out of 100. Here are the marks for 60 students.

```
71  62  40  72  59  63  43  81  44  23
55  52  55  58  66  31  45  54  57  59
63  61  54  42  35  47  33  62  41  73
57  82  26  71  52  48  38  65  52  56
68  36  49  63  57  53  77  65  27  88
41  62  35  47  63  39  62  43  46  51
```

(a) Copy and complete the frequency table to show the students' marks.

Mark	Tally	Frequency
20–29		
30–39		
40–49		
50–59		
60–69		
70–79		
80–89		
Total		

(b) The pass mark for this test was 40 out of 100. How many students passed the test?

2 A newsagent recorded the number of newspapers sold on each day in January:

```
40  62  67  40  49  52  57  42
46  44  48  55  53  51  56  58
58  59  60  44  52  63  48  49
42  53  57  56  53  61  51
```

(a) Draw up and complete a frequency table, using class intervals 40–44, 45–49, and so on.

(b) In order to cut costs, the newsagent decides that he will stock only 60 newspapers each day. In January, on how many days would he have sold out of newspapers?

3 Here are the ages of 60 guests at a wedding.

```
30  57   4  40  32  63  27  64  38  12
66  36   2  35  62  25  66  48  33  14
61  34  63   6  39  68  45  67  22   9
32  33   3  42  72  35  34   5  77  55
31  79  40   7  37  18  74  37  29  32
 2  38   1  32   5  33   7   3  26   5
```

(a) Design and complete a frequency table with equal class intervals to sort these data.

> Divide the data into 8 classes.

(b) Which age group is the most common?

(c) The bride's mother was heard to say, 'Most of the children at the wedding seem to be quite young.'
Explain whether you think she was correct.

4 Fifty people were each asked to name their favourite pastime. Here are their choices:

> Qualitative data can also be put into grouped frequency tables.

football; weeding; cricket; reading; growing vegetables; walking; tennis; mowing the lawn; sleeping; writing; travelling abroad; watching TV; hanging wallpaper; playing the clarinet; theatre; planting bulbs; painting and decorating; camping; football; cinema; weeding; netball; mowing the lawn; tennis; going away for the weekend; growing vegetables; lying in bed; cricket; making furniture; roller hockey; music concerts; planting bulbs; football; reading; sleeping; writing; cinema; singing; playing the violin; football; reading; weeding; travelling abroad; mowing the lawn; sleeping; planting bulbs; painting and decorating; camping; reading; singing.

(a) Design and complete a grouped frequency table. Your classes could be *music, sport, gardening, relaxing, film and theatre, DIY, holidays* and *reading and writing*.

(b) Were there any pastimes that you found difficult to classify? Explain why.

5 Forty people took part in a competition flying model aeroplanes.
The competitors included experts and beginners.
Each competitor was given a score out of 120.
Here are the scores awarded.

111	97	36	41	115	15	112	99	56	105
73	71	47	33	46	105	109	22	56	52
109	43	36	95	17	48	85	107	42	35
56	28	103	59	57	116	38	29	53	61

(a) Draw up a frequency table, using class intervals of 1–10, 11–20, and so on.

(b) Using the same data, draw up a new frequency table, using class intervals of 1–20, 21–40, and so on.

Hint: Combine groups from the first frequency table.

(c) Using the same data, draw up a third frequency table, using class intervals of 1–40, 41–80 and 81–120.

(d) Which of the three frequency tables gives the best information about the distribution of scores? Explain your answer.

(e) Explain the limitations of the other two frequency tables.

3.3 Classes of varying width and open-ended classes

Data do not always have to be grouped in equal class intervals.
The distribution of data is often easier to see if class intervals are varied appropriately.

Example 3

In a competition, students were loaded up with exercise books until they dropped one to the floor.
The raw data below show the number of exercise books balanced by the first 50 students who tried.

Sort these data into groups.

7	37	41	33	44	31	49	15	32	40
69	45	42	40	38	52	27	42	31	37
36	31	39	53	46	82	41	47	43	42
38	39	40	43	41	37	32	51	11	33
32	66	45	42	39	35	47	58	3	44

Although there is a large range, there are few data below 30 or over 50. Hence, wider class intervals would be more appropriate at the extremes of the data.
By observation, it is clear that most of the data are grouped between 30 and 49. It would be more useful if there were smaller class intervals in this range.

This frequency table has class intervals of different sizes to suit the data.

Number of exercise books	Frequency
0–29	5
30–34	8
35–39	10
40–44	14
45–49	6
50–59	4
60–	3

More students are going to compete. Before these extra data are collected, there is no way of knowing the highest score. For this reason, the last group has been left **open**. The class has no **upper bound**.

You learnt about bounds in Section 2.8, page 24.

■ **When there is not an even spread of data across the range use class intervals of varying width.**

■ **When the extreme values of the data are not known, leave the first and/or last interval open.**

Exercise 3C

1 Henry asked 30 students in his class how much money they had in their pocket at the beginning of a statistics lesson.
  The amounts are shown below.

  £1.20 £2.43 76p   0p   £1.63 £1.42 £2.09 £1.80
  £1.36 £10.50 37p   £1.28 £2.61 £1.60 £1.50 £2.00
  £1.22 £1.55 £3.50 £2.32 £1.40 £1.50 £2.00 £1.87
  £1.75 £2.50 £1.35 £1.40 £1.59 £2.05

  Copy and complete the frequency table.

Amount	Tally	Frequency
£0–£1.00		
£1.01–£1.50		
£1.51–£2.00		
£2.01–£3.00		
£3.01–		
Total		

2 Here are the ages of 50 people at a night club.

```
30  24  15  31  23  28  32  29  33  48
31  37  42  18  20  34  40  25  36  31
29  32  26  33  25  27  32  22  29  28
21  35  34  29  30  34  26  32  22  31
29  35  19  28  24  33  27  59  32  27
```

(a) What is the age of the youngest person?

(b) What is the age of the oldest person?

(c) Between which two ages do most of the people fall?

(d) Design and complete a frequency table for these data. Use class intervals of different sizes. ———————————— Use a lower class of ≤20 and an upper class of ≥46.

3 Eighty people were asked how many television programmes they had watched in a week. The results of the survey are shown in two different frequency tables.

Table 1 Equal class intervals

Number of programmes	Frequency
0–9	4
10–19	54
20–29	16
30–39	4
40–49	1
50–59	0
60–69	1
Total	80

Table 2 Varied class intervals

Number of programmes	Frequency
0–8	3
9–12	21
13–16	8
17–20	32
21–30	11
31–40	3
41–	2
Total	80

(a) Write down two limitations of table 1.

(b) In table 2, why has the last class been left open?

(c) Only one person watched over 60 programmes in a week. What reason could there be that this does not happen more often?

(d) What extra information can you read from table 2 that was hidden in table 1?

4 Sue and John decided to conduct a survey into the ages of 125 people at a classical music concert. Both collected the same data.
Their frequency tables are shown on the next page.

Sue's frequency table

Age	Frequency
0–9	1
10–19	0
20–29	2
30–39	55
40–49	56
50–59	8
60–69	2
70–79	0
80–89	1
Total	125

John's frequency table

Age	Frequency
0–29	3
30–34	18
35–39	37
40–44	43
45–49	13
50–59	8
60–	3
Total	125

(a) What is the main difference between the two frequency tables?

(b) Explain why John has made such a wide class interval for people below the age of 30.

(c) Which frequency table shows more detail about the most common age ranges? Explain your answer.

(d) Why did John leave the last interval open when collecting these data?

(e) Write down two ways in which Sue could have improved her frequency table.

3.4 Two-way tables

■ **A two-way table lets you show two variables at the same time.**

This is called bivariate data, see page 5.

Example 4

The two-way table below shows some information about the gender of students in Year 11 at Hitchen High School. Copy and complete the table, by filling in the missing information.

The rows classify each student by gender.

Each column represents a Year 11 class.

There are 61 boys altogether in Year 11.

	11A	11B	11C	11D	Total
Boys	18	16		14	61
Girls	12		14	19	
Total	30	33	27	33	

There are 12 girls in form 11A.

There are 33 students in form 11D.

You can use the information provided to calculate the missing values.

For example, how many girls are there in form 11B?
The column for 11B shows that there are 33 students in 11B, 16 of whom are boys.

$$33 - 16 = 17 \text{ girls}$$

The completed table will look like this:

	11A	11B	11C	11D	Total
Boys	18	16	13	14	61
Girls	12	17	14	19	62
Total	30	33	27	33	123

Column – 11B	Column – 11C	Row – Girls	Column total
$33 - 16 = 17$	$27 - 14 = 13$	$12 + 17 + 14 + 19 = 62$	$61 + 62 = 123$

■ **Organising data into a two-way table can help you to calculate missing information.**

Using ICT for two-way tables

When you wish to use the same two-way table several times with different numerical values, it is useful to use formulae on a spreadsheet.

Example 5

Lisa decided to record information about the guests at her aunt's wedding.
She began to record her findings on a spreadsheet.

	A	B	C	D	E	F
1		Adult male	Adult female	Boy	Girl	Total
2	Bride's family	17	22	8	5	
3	Groom's family	13	25	6	9	
4	Friends	18	15	3	2	
5	Total					
6						
7						

(a) Copy this table onto a spreadsheet of your own.

(b) Use spreadsheet formulae to fill in the total column and the total row.

(a) Enter the information into a new spreadsheet.

- In cell B1, type 'Adult male', in C1 type 'Adult female'. Continue until you have entered all the information shown above.

If the information does not fit into some cells, then you may need to make some columns wider.
To make column A wider:

- Move the cursor to the right hand edge of the cell marked 'A'.

- Hold down the left button on the mouse, and drag the mouse to the right until the column is the correct width.

Repeat the same process with the other columns.

(b) To find the total number of adult males, you need to add up the numbers in cells B2, B3 and B4.

- In cell B5, type '=SUM(B2:B4)'.
  This works out 'the sum of cells B2 to B4'.

- In cell C5, type '=SUM(C2:C4)'. This gives the total number of adult females.

- In cell D5, type '=SUM(D2:D4)'. This gives the total number of boys.

- In cell E5, type '=SUM(E2:E4)'. This gives the total number of girls.

- In cell F2, type '=SUM(B2:E2)'. This gives the total number of people in the bride's family.

- In cell F3, type '=SUM(B3:E3)'. This gives the total number of people in the groom's family.

- In cell F4, type '=SUM(B4:E4)'. This gives the total number of friends.

- In cell F5, type '=SUM(B5:E5)'. This gives the total number of people at the wedding.

Exercise 3D

1 Charles asked 87 adults and children whether they were right-handed or left-handed.

	Adults	Children	Total
Right-handed	32		
Left-handed		22	
Total	47		87

(a) Copy and complete the two-way table.

(b) How many right-handed children were in the sample?

Charles thinks that it is more likely for a child to be left-handed, than it is for an adult.

(c) Do these data support his view?

2   Copy and complete the two-way table to show drinks chosen by 28 children at a birthday party.

	Lemonade	Orange juice	Total
Girls		9	
Boys	10	6	
Total			

Hint: Fill this one first!

3   Victoria conducted an experiment to see whether or not a piece of buttered toast was more likely to land 'butter-side down'. Each time, she either dropped the toast or threw it.
   ● She conducted the experiment 82 times in total.
   ● She dropped the toast 37 times in total.
   ● When the toast was thrown, it landed 'butter-side down' on 21 occasions.
   ● When the toast was dropped, it landed 'butter-side up' on 11 occasions.

   (a) Copy and complete the two-way table.

	'Butter-side down'	'Butter-side up'	Total
Dropped			
Thrown			
Total			

   (b) In the experiment, which way up did the toast land most frequently?

4   Members of a fitness club can either pay full membership, or pay only for the weekends. Under-14s pay half-price.

   Only 12% of the members are under 14 and pay full membership. 29% of the members pay for weekends only. 63% of the members are over 14.
   Use a two-way table to find out what percentage of over-14s pay full membership.

Hint: There is 100% altogether!

5 In a survey, teachers of different subjects were asked how they preferred to travel to work. Some of the results are shown in the table below.

	Car	Bus	Cycle	Walk	Other	Total
English		4	0	1	0	
Games	3	1	18		3	32
Geography	8		1	18		32
Maths	28	3	1		1	
Science	16	5	7	6		
Total		17		33	9	156

(a) Copy and complete the two-way table.

(b) How many science teachers were involved in the survey?

(c) In total, how many teachers travelled by car?

(d) How many maths teachers preferred to walk to school?

6 Jenny looked at results from the football season 2001/2002 to see whether teams playing in red were more successful than those playing in blue.
She began to record her findings on a spreadsheet.

	A	B	C	D	E
1		Win	Lose	Draw	Total
2	Team playing in blue	367	185	229	
3	Team playing in red	442	229	255	
4	Total				
5					
6					
7					

(a) Copy this table onto a spreadsheet of your own.

(b) Use spreadsheet formulae to fill in the total column and the total row.

7 The spreadsheet below shows the amount of money (in pounds) taken each day by a charity shop in the first four weeks of the year.

	A	B	C	D	E	F
1		Week one	Week two	Week three	Week four	Total
2	Monday	367	185	410	229	
3	Tuesday	442	229	546	255	
4	Wednesday	552	338	670	310	
5	Thursday	387	286	412	272	
6	Friday	298	279	359	183	
7	Saturday	646	550	812	435	
8	Total					

(a) Copy this table onto a spreadsheet of your own.

(b) Use spreadsheet formulae to fill in the column and the total row.

(c) How much money was taken in total, on the first four Saturdays of the year?

(d) How much money was taken in total during week two?

(e) The amount for the Tuesday of week two should say '292'. Change this figure in your spreadsheet. What is the new total for week two?

3.5 Other tables and databases

There are many other types of tables that are used to display a variety of data. Examples include distance tables, train timetables, calendars, price lists, etc.

■ **A database is a collection of information. Computers can store huge databases that can be easily selected, sorted and ordered at the touch of a button.**

■ **A summary table shows data that have been sorted and summarised, and is easier to interpret than the original raw data.**

The following exercise shows a variety of tables that you may come across in your everyday life.

Exercise 3E

1

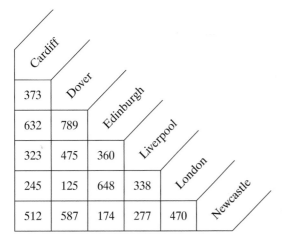

In the table, all distances are given in kilometres.

(a) What is the distance between:
  (i) Liverpool and Dover,
  (ii) Newcastle and Edinburgh,
  (iii) London and Cardiff?

(b) Mrs McQueen travels from Dover to London, and then on to Newcastle. How far does she travel altogether?

(c) Mr King is a travelling salesman. In one week he travelled from London to Cardiff, to Liverpool, to Newcastle and finally returned to London. How far did he travel in that week?

2 The diagram shows a section of a train timetable.

Train timetable – Andwich to Elchester									
	X			**X**	**S**				**S**
Andwich	07:30	08:45	10:40	12:55	13:15	15:30	17:25	19:50	23:30
Balstone	07:42	08:57	10:52	13:07	13:27	15:42	17:37	20:02	23:42
Ciffingham	07:58	09:13	11:08	13:23	13:43	15:58	17:53	20:18	23:58
Dilsbury	08:09	09:24	11:19	13:34	13:54	16:09	18:04	20:29	00:09
Elchester	08:20	09:35	11:30	13:45	14:05	16:20	18:15	20:40	00:20

X service not available on Saturday
S Saturday only

(a) What time is the earliest train from Andwich to Elchester on a Monday?

(b) What time is the earliest train from Andwich to Elchester on a Saturday?

(c) Winston catches the train from Ciffingham at 11 08. What time does he arrive at Dilsbury?

(d) Cathy catches the train from Balstone at 20 02. What time does she arrive at Elchester?

(e) How long does the train take to travel between
  (i) Andwich and Balstone,
 (ii) Balstone and Ciffingham,
(iii) Ciffingham and Dilsbury,
 (iv) Andwich and Elchester?

(f) Daniel arrives at Dilsbury at 13 30 on a Saturday. What time is the next train to Elchester?

3 The cost of car insurance is dependent on many factors. These include the gender of the driver, and the area in which the driver lives.
The table shows the cost of insurance for a particular make and size of car.

Age	Gender	Area				
		A	B	C	D	E
17–25	M	£484	£366	£633	£500	£558
	F	£387	£293	£506	£400	£446
26–35	M	£397	£300	£519	£410	£458
	F	£315	£238	£411	£325	£363
36–50	M	£242	£183	£317	£250	£279
	F	£194	£146	£253	£200	£223
51+	M	£266	£201	£348	£275	£307
	F	£266	£201	£348	£275	£307

(a) How much would the insurance cost for:
   (i) Arthur, a 42-year-old male who lives in area C?
   (ii) Amrita, a 20-year-old female who lives in area E?
   (iii) Michael, a 70-year-old male from area A?
(b) In which age group do males and females pay the same for their insurance?
(c) In area B, how much more would insurance for a 33-year-old cost for a male than a female?
(d) Describe a person who pays the most for insurance according to this table.
(e) Describe a person who pays the least for insurance according to this table.

4 The summary table below shows eight hotels and the facilities they offer.

	Evening meal	Swimming pool	Bicycle hire	Fitness suite	Bar	Laundry	Room service	Wake-up call	Satellite TV	Kid's play area
The HILLSTONE	✓	✓	✓		✓	✓		✓		✓
FIVEWAYS	✓		✓		✓	✓		✓	✓	
ROLLING HILLS	✓		✓		✓				✓	
PORTENDALES	✓	✓		✓		✓	✓	✓		
The MARION	✓	✓	✓		✓			✓	✓	
The TOWN	✓	✓			✓	✓	✓	✓		✓
WALKERSTONES	✓	✓	✓		✓		✓			
The RED TIGER	✓				✓			✓	✓	✓

(a) Which two hotels have a fitness suite?

(b) Which hotel has both a kids' play area and bicycle hire?

(c) Which hotels could you go to if you wanted both a wake-up call and room service?

(d) Raj enjoys swimming and cycling and likes to have a wake-up call in the morning. Which hotels would you recommend?

(e) Which hotel does not have a bar?

5 Ask 10 people from your class for the following data:
- the month of their birthday,
- their height in centimetres,
- their gender,
- their shoe size.

Design a summary table to show these data.

3.6 Pictograms

■ **Diagrams and charts are used to represent data in a clear and simple way, but do not always show the exact information.**

■ **A pictogram uses symbols to represent a certain number of items.**

Example 6

This pictogram shows the number of computers in each of four high schools.

(a) Which school has the most computers?

(b) How many computers can be found at Woodridge High School?

(c) Estimate the number of computers at Hursley Comprehensive.

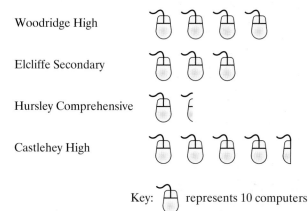

Key: represents 10 computers

(a) In this example it is immediately clear that Castlehey has the most computers.

(b) Woodridge has 4 complete symbols, each representing 10 computers. It is clear that Woodridge has exactly 40 computers.

(c) Some of the detail has been lost. It is not possible to say exactly how many computers the part symbols stand for. We can estimate that Hursley has about 13 computers, as there is less than one and a half symbols.

When drawing a pictogram:

- Draw each symbol exactly the same size.
- Make sure that the symbols are spread out in the same way in each row.
- Include a clear key.

Exercise 3F

1 Woodridge High School organised an activity day. The pictogram shows how the pupils decided to spend their day.

Key: represents 50 pupils

(a) Which activity was the most popular?

(b) How many pupils went bowling?

(c) Estimate how many pupils went to the theme park.

(d) Estimate how many pupils took part in the least popular activity.

(e) Approximately how many pupils took advantage of the activities offered?

2 Forty customers at a supermarket were asked which of the local towns they came from. Their responses are shown in the frequency table. Draw a pictogram to display this information.

Town	Frequency
Upshaw	5
Bunton	10
Chuckleswade	18
Newtown	5
Shenford	2

3 The table below shows the amount of money that is spent on Education, Health, Transport and Emergency services in the city of Suncastle.

Area of spending	Amount of money
Health	£57 000 000
Education	£62 000 000
Transport	£15 000 000
Emergency services	£34 000 000

Draw a pictogram to display this information.

The next question looks at an example of how badly drawn diagrams can be misleading.

4 Four fishermen kept a record of how many fish they had caught during a year.
They tried to show their results in the pictogram below.

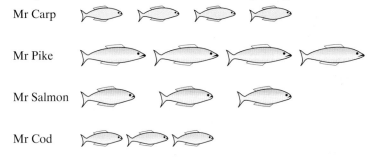

Mr Carp

Mr Pike

Mr Salmon

Mr Cod

(a) Explain why it is impossible to tell how many fish were caught by each fisherman.

(b) Mr Pike uses this pictogram to show that he caught the most fish. Explain why this is misleading.

(c) Mr Salmon caught the same number of fish as another fisherman. Who was this?
Explain why the pictogram can mislead you in this question.

3.7 Bar charts and vertical line graphs

■ **Bar charts are a simple way to show trends for discrete data. The bars may be horizontal or vertical.**

Example 7

This bar chart shows the marks out of ten scored by a class in a mental maths test.

(a) What is the range of marks in this mental test?

(b) How many pupils scored full marks?

(c) What is the most common mark?

(d) How many pupils are in the class?

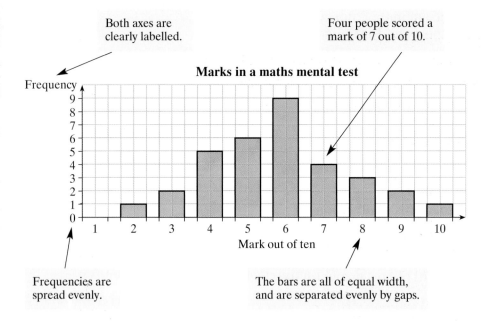

Both axes are clearly labelled.

Four people scored a mark of 7 out of 10.

Frequencies are spread evenly.

The bars are all of equal width, and are separated evenly by gaps.

The bar chart above shows exact frequencies for a small amount of data.

(a) A quick glance shows that the marks ranged from 2 to 10.

(b) Only one pupil scored 10 out of 10 (full marks).

(c) The most common mark was clearly 6 out of 10, as this has the highest bar.

(d) Find the total number of children in the class by adding the frequencies represented by each bar:

$$0 + 1 + 2 + 5 + 6 + 9 + 4 + 3 + 2 + 1 = 33$$

Bar charts can be used to show approximate frequencies for a large amount of data.

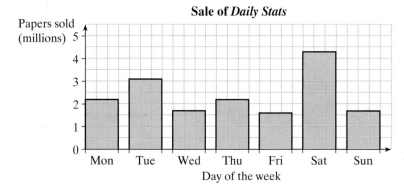

This bar chart shows the number of copies of the *Daily Stats* sold in the UK in one week.

It is not possible to show exactly how many newspapers were sold on each day. You can estimate that 2.2 million newspapers were sold on Monday.

When drawing a bar chart:
- label both axes clearly and write a title,
- make all bars the same width and separate them evenly with gaps,
- design the frequency scale so that the chart is a sensible size.

A bar chart can also be used to show grouped discrete data.

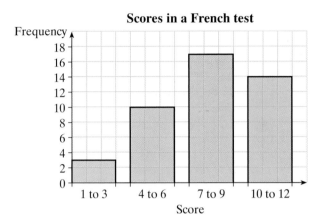

Scores in a French test

Write each class interval directly beneath the correct bar.

Vertical line graphs

■ **A vertical line graph can be used to display discrete data. The format is very similar to a bar chart, with a series of vertical lines evenly spread.**

Example 8

The vertical line graph shows the number of people in each car that passed the gates of a school between 9 00 a.m. and 10 00 a.m.

(a) Which number of people occurred most frequently?

(b) Estimate the number of cars which contained 2 people.

(c) Why do you think there were no cars with more than 5 people?

(d) Why might these results not be entirely reliable?

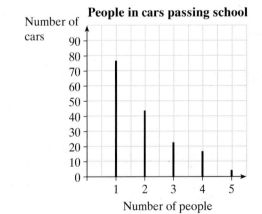

People in cars passing school

(a) 1 person in a car occurred most frequently.

(b) About 44 cars contained 2 people.

(c) There are not many cars that can hold more than 5 people.

(d) It is difficult to count the number of passengers in a car as it passes.

Exercise 3G

1   The vertical line graph shows the shoe sizes of students in class 11G.

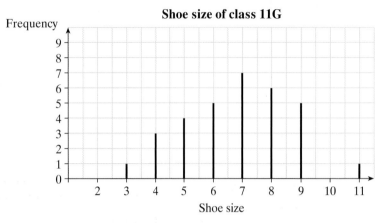

Shoe size of class 11G

(a) How many students had size 5 shoes?

(b) What is the most popular shoe size?

(c) Which size of shoe does the student with the smallest feet wear?

(d) Which two shoe sizes have identical frequencies?

(e) How many students were in 11G?

(f) Construct a pictogram to show the same information.

2   The bar chart shows the number of pets owned by members of class 11H.

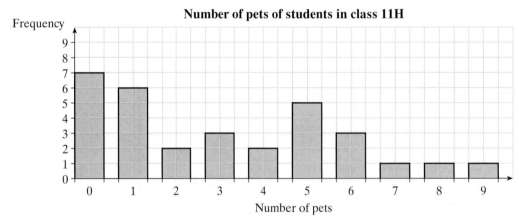

Number of pets of students in class 11H

(a) How many students did not have any pets?

(b) What is the largest number of pets owned by a member of this class?

(c) How many students are in 11H altogether?

(d) A student is chosen at random from the class. How many pets is he most likely to have?

(e) Is this a good way to show these data? Explain your answer.

3 The frequency table shows the number of children in 30 families.

Number of children	0	1	2	3	4	5
Frequency	4	5	14	4	2	1

(a) Draw a pictogram to display this information, using suitable symbols.
(b) Draw a bar chart to display this information.
(c) Which of these charts did you find easiest to draw? Explain your answer.
(d) Which chart gives more accurate information? Explain your answer.

4 Forty customers at a supermarket were asked which of the local towns they came from. Their responses are shown in the frequency table.

Town	Frequency
Upshaw	5
Bunton	10
Chuckleswade	18
Newtown	5
Shenford	2

(a) Draw a bar chart to display this information
(b) Which of the towns is most likely to be nearest to the supermarket?
(c) In question 2 of Exercise 3F, these data were used to draw a pictogram. Which type of presentation do you prefer? Explain your answer.

5 A book, *The History of Mathematics*, was opened at random and the lengths of 1000 words were counted. The results were as follows.

Number of letters	1	2	3	4	5	6	7	8	9	10
Frequency	35	132	306	183	123	96	62	41	14	8

(a) Draw a pictogram to display this information.
(b) Is the pictogram a good way to show these data? Explain your answer.
(c) Draw either a bar chart or vertical line graph for these data. Explain your choice.

6 Conduct a survey to find out the number of mobile phones owned by the families of your classmates. Choose a frequency diagram to show this information. Explain your choice.

The next question looks at an example of how badly drawn diagrams can be misleading.

7 Ninety-five people were asked which team they preferred; Rovers or City.
Andrea, Brian, Carin and Des each drew a bar chart to display this information.

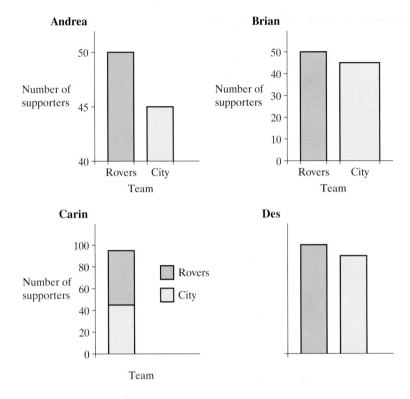

(a) Andrea is a Rovers supporter.
   She said 'From my bar chart there seem to be twice as many
   Rovers supporters.'
   Explain why Andrea's bar chart is misleading.

(b) Brian is a City supporter.
   Explain how Brian's bar chart is biased in favour of City.

(c) Carin said, 'Rovers and City seem to have about the same
   number of supporters.'
   Explain why Carin's choice of bar chart has made it difficult
   to compare frequencies.

(d) Des cannot find out anything from his bar chart. Explain why.

3.8 Using bar charts to make comparisons

Multiple bar charts

■ **Multiple bar charts have more than one bar for each class.**

Example 9

Janet asked 40 boys and 40 girls their shoe size.
The results are shown on the **multiple bar chart** below.

(a) What is the most common shoe size for girls?

(b) Which gender wore more size 8 shoes?

(c) Which shoe size was worn by an equal number of both boys and girls?

(d) Only one student had size 2 shoes. Was this a boy or a girl?

(e) Janet thinks that boys have larger feet than girls. Do you think that she is correct? Explain your answer.

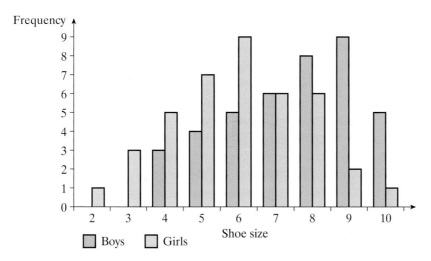

(a) The green bars represent the shoe sizes of the girls. The most common shoe size for girls is size 6 – represented by the highest green bar.

(b) Boys had more size 8 shoes than girls – the red bar shows a higher frequency.

(c) Equal numbers of boys and girls wear size 7 shoes – the two bars are the same height.

(d) One girl wore size 2 shoes.

(e) The bar chart shows that girls' shoes range from size 2 to size 10, but the most common shoe sizes for the girls were around 5 or 6. The red bars show that boy's shoe size ranges from size 4 to size 10, with sizes 8 and 9 being the most common.
The bar chart clearly shows that more boys have the larger sizes. Janet's theory would seem to be correct, but only if she asked boys and girls of similar ages.

Compound bar charts

The same information can be shown on a **compound bar chart**.

■ **A compound bar chart has single bars split into separate sections for each category. You can compare groups, and you can see the distribution of the data as a whole.**

This compound bar chart shows the boys' frequencies *on top* of the girls' frequencies.

It is easy to compare the boys' and girls' shoe sizes, but it is now more difficult to see the boys' frequency distribution.

The benefit of this chart is that you can see the distribution of shoe sizes for all the students.

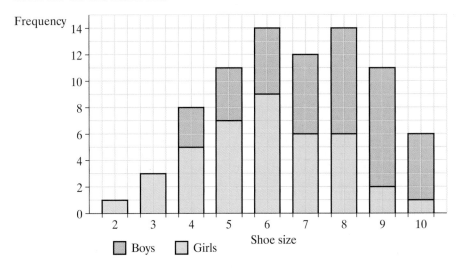

Exercise 3H

1   Paul sells scarves in three colours. Each scarf is black or red or blue. The multiple bar chart below shows his sales over one week.

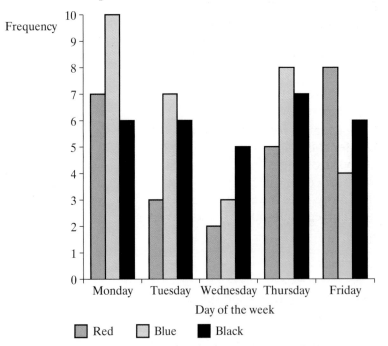

(a) What colour scarf was most popular on Monday?
(b) Which colour had the most consistent sales over the whole week?
(c) How many scarves were sold on Wednesday?
(d) How many red scarves were sold over the whole week?
(e) On which day did Paul sell the fewest scarves?

2   Roland owns a market stall, from which he sells fruit and vegetables. One day he compared the types of fruit that he sold in the morning and afternoon.
His results are shown in the compound bar chart.

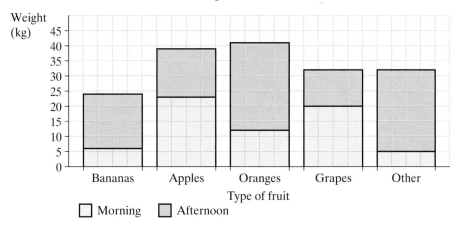

(a) Which fruit was the most popular overall?

(b) Which fruits were more popular in the morning than the afternoon?

(c) Estimate the weight of apples sold in the morning.

(d) Estimate the weight of grapes sold in the afternoon.

(e) Estimate the weight of bananas sold in the whole day.

3   (a) Use the data in question 1 to construct a compound bar chart.

(b) Use the data in question 2 to construct a multiple bar chart.

4   Conduct a survey of your class to find the number of children in their families. Draw a bar chart that can be used to compare the family size of the boys and girls in your class.

3.9 Using ICT to draw bar charts

Example 10

The spreadsheet below shows information about the total number of police arrests in Clarksville in the first 4 months of this year.

(a) Copy this table onto a spreadsheet of your own.

(b) Use the spreadsheet to draw a bar chart.

	A	B
1	Month	Arrests in Clarksville
2	January	193
3	February	132
4	March	88
5	April	112

(a) Enter the information into a new spreadsheet.
- In cell A1, type 'Month', in B1 type 'Arrests in Clarksville'.
- Continue until you have entered all the information shown above.

If the information does not fit into some cells, then you may need to make some columns wider.
To make column A wider:
- Move the cursor to the right hand edge of the cell marked 'A'.
- Hold down the left button on the mouse, and drag the mouse to the right until the column is the correct width.

Repeat the same process with column B.

(b)
- Move the cursor to cell A1.
- Hold down the left button on the mouse, and drag the mouse to cell B5. All ten cells will now be highlighted.
- Click **Insert** on the toolbar, then **Chart**.
- Select **Column**, as your type of chart, and select the picture that matches the chart you require.
  The chart should look like the one below.

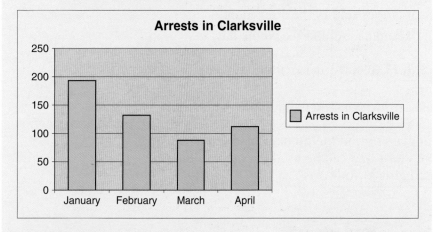

The title is added automatically by Excel.

Example 11

The spreadsheet below shows information about the total number of police arrests in the three towns of Clarksville, Boystown and Rowchester in the first 4 months of this year.

	A	B	C	D
1	Month	Arrests in Clarksville	Arrests in Boystown	Arrests in Rowchester
2	January	193	88	152
3	February	132	110	75
4	March	88	135	38
5	April	112	55	24

(a) Copy this table onto a spreadsheet of your own.

(b) Use the spreadsheet to draw a multiple bar chart.

(c) Use the information to draw a compound bar chart.

(a) Enter the information into a new spreadsheet, as in Example 10.

(b) and (c)
  - Move the cursor to cell A1.
  - Hold down the left button on the mouse and drag the mouse to cell D5. All 20 cells will now be highlighted.
  - Click **Insert** on the toolbar, then **Chart**.
  - Select **Column**, as your type of chart, and select the picture that matches the chart you require.

The charts should look like the ones below.

(b)

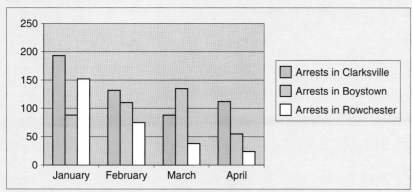

(c)

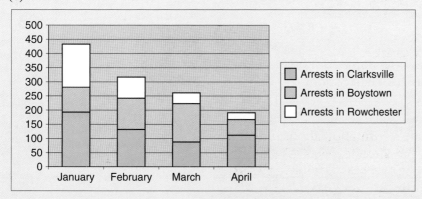

Exercise 3I

1   Debbie asked 40 adults to choose their favourite polygon, and then classified their answers by the number of sides. The results are shown on the section of a spreadsheet below.

(a) Copy this table onto a spreadsheet of your own.

(b) Use your spreadsheet to generate a bar chart for these data.

	A	B
1	Type of shape	Frequency
2	Triangle	9
3	Quadrilateral	13
4	Hexagon	8
5	Octagon	6
6	Other	4
7	Total	40

2 Forty customers at a supermarket were asked which of the local towns they came from.
Their responses are shown in the frequency table.

Town	Frequency
Upshaw	5
Bunton	10
Chuckleswade	18
Newtown	5
Shenford	2

(a) Transfer this information onto a new spreadsheet.

(b) Use your spreadsheet to generate a bar chart for these data.

(c) In question 4 of Exercise 3G, these data were used to draw a bar chart by hand. Write down one advantage and one disadvantage of using a spreadsheet to construct a bar chart for these data.

3 The diagram below shows a section of a spreadsheet.
The spreadsheet displays the number of boys and girls in each year group at Dunkley High School.

	A	B	C	D
1		Boys	Girls	Total
2	Year 7	103	65	168
3	Year 8	76	77	153
4	Year 9	78	92	170
5	Year 10	112	98	210
6	Year 11	99	80	179
7	Total	468	412	880

Be careful. You do not want to include the 'total' column in your bar chart.

(a) Copy this table onto a spreadsheet of your own.

(b) Use the spreadsheet to generate either a multiple bar chart or a compound bar chart to show these data.

3.10 Pie charts

■ **A pie chart is a good way of displaying data when you want to show how something is shared or divided**

Each sector of this pie chart represents one of the types of houses. You cannot easily read the exact frequencies, but the *proportion* of each type of housing is clear.

Types of housing in Showtown

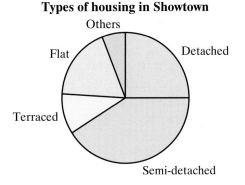

You can see that the most common type of house is semi-detached. You cannot work out how many houses there are without being given more information.

Example 12

Draw a pie chart to show what 24 people in a hotel ordered for breakfast.

Choice of breakfast	Frequency
Cereal	6
Full English	11
Continental	5
Fruit	2
Total	24

First calculate the angles at the centre of the circle for each sector. The angles at the centre of the pie chart will add up to 360°.

There are two different methods for calculating the angles.

Method 1
Work out what fraction of the guests had each type of breakfast. For example, $\frac{6}{24}$ of the guests chose cereal for their breakfast.

Cereal: $\dfrac{6}{24} \times 360° = 90°$

Full English: $\dfrac{11}{24} \times 360° = 165°$

Continental: $\dfrac{5}{24} \times 360° = 75°$

Fruit: $\dfrac{2}{24} \times 360° = 30°$

Before you try to draw the pie chart, check that the angles total 360°:

$90° + 165° + 75° + 30° = 360°$

Method 2 – the unitary method
Work out how many degrees are represented by each guest.

24 guests are represented by 360°.

$360° \div 24 = 15°$, so each guest is represented by 15°.

Cereal: $6 \times 15° = 90°$
Full English: $11 \times 15° = 165°$
Continental: $5 \times 15° = 75°$
Fruit: $2 \times 15° = 30°$

To draw a pie chart, begin by drawing a circle. Make sure that you mark the centre clearly.

If you are drawing only one pie chart, then you can choose the size of the radius, probably between 5 cm and 8 cm.

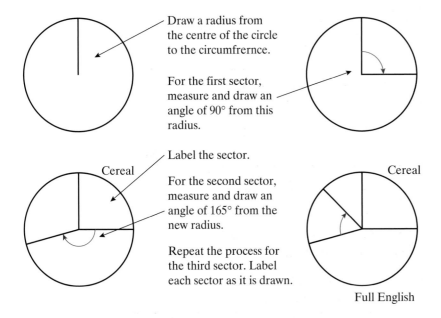

Draw a radius from the centre of the circle to the circumfrernce.

For the first sector, measure and draw an angle of 90° from this radius.

Label the sector.

For the second sector, measure and draw an angle of 165° from the new radius.

Repeat the process for the third sector. Label each sector as it is drawn.

The final sector should already be drawn. Measure it to check that it equals 30°.

When you have completed the pie chart, shade the sectors to make the proportions clearer. The sectors can be labelled inside, outside, or using a key.

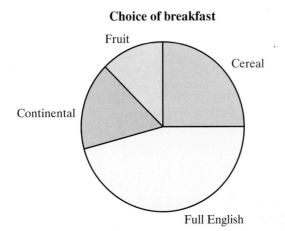

Choice of breakfast

Exercise 3J

1 90 boys were asked their favourite sport.

Sport	Football	Rugby	Cricket	Hockey	Other
Frequency	42	12	21	10	5

Draw a pie chart to display this information.

2 The table shows the length of time spent on different items on a radio station one day last week.

Item	News	Adverts	Music	Chat	Weather
Time (hours)	2	4	14	3	1

(a) Calculate the angle in a pie chart that would represent Music.
(b) Draw a pie chart to display this information.

3 60 cars in a car park are surveyed to find the place of their manufacture.

Country	France	Germany	Japan	Italy	UK
Frequency	14	7	13	8	18

(a) What fraction of the cars was manufactured in Germany?
(b) Draw a pie chart to display this information.

4 7200 people voted for their favourite act in a TV talent competition. The table shows the results of their votes.

Act	Duet	Acrobat	Juggler	Comedian	Snake Charmer
Votes	200	900	1800	3100	1200

(a) Which act was the least popular?
(b) What fraction of the votes was scored by the winning act?
(c) Draw a pie chart to display this information.

5 Joanne spent £120 on a shopping trip. The table below shows how she spent her money.

Items	Amount (£)
Food and drink	14
Clothes	63
CDs	16
Presents	
Total	120

(a) How much money did Joanne spend on presents?
(b) Draw a pie chart to display this information.

6 A survey was conducted of the number of children in each family in Fromley.
Some of the percentages are shown in the table.

Number of children in family	Amount
0	15%
1	20%
2	35%
3	25%
more than 3	

(a) What percentage of families had more than 3 children?

(b) Draw a pie chart to display this information.

3.11 Pie charts for real data

■ **When constructing pie charts for real data, you will often need to use rounded values for the size of the angles.**

Example 13

John asked all 224 pupils in his year group about their favourite subject. The results are shown in the table.

Subject	Maths	English	Art	Science	P.E.	Other
Frequency	59	38	52	41	16	18

Draw a pie chart to show these data.

Method 1

Maths $\quad \dfrac{59}{224} \times 360 = 94.8°$

This has been rounded to 1 decimal place, but with a protractor it is difficult to measure more accurately than the nearest degree.

English $\quad \dfrac{38}{224} \times 360 = 61.1°$

Art $\quad \dfrac{52}{224} \times 360 = 83.6°$

Science $\quad \dfrac{41}{224} \times 360 = 65.9°$

P.E. $\quad \dfrac{16}{224} \times 360 = 25.7°$

Other $\quad \dfrac{18}{224} \times 360 = 28.9°$

You will now need to think carefully about how you round these figures, to ensure that they add up to 360.
For example, try rounding 83.6 and 25.7 down to 83.5 and 25.5.
It is just possible to measure to half a degree.

Rounding these figures to the nearest degree gives 95, 61, 84, 66, 26, 29 but these total 361 degrees.

This is because so many of the figures have been rounded up. You need to think carefully about how you round these figures so that they total 360.

Subject	Maths	English	Art	Science	P.E.	Other
Angle	95	61	83.5	66	25.5	29

Method 2
Work out how many degrees represent each pupil.

$$360 \div 224 = 1.607\,142\,857\ldots$$

It is important that you do not round this number at this stage. Use the memory of your calculator to store this figure.

$1.607\,142\,857 \times 59 = 94.8°$ $1.607\,142\,857 \times 41 = 65.9°$
$1.607\,142\,857 \times 38 = 61.1°$ $1.607\,142\,857 \times 16 = 25.7°$
$1.607\,142\,857 \times 52 = 83.6°$ $1.607\,142\,857 \times 18 = 28.9°$

Now round off in the same way as the first method.

Using ICT to draw pie charts

Example 14

The spreadsheet shows the composition of the earth's atmosphere.

(a) Copy this data onto a spreadsheet of your own.

(b) Use your spreadsheet to draw a pie chart.

	A	B
1	Gas	Percentage of Earth's atmosphere
2	Oxygen	21
3	Nitrogen	78
4	Other (including argon)	1

(a) Enter the information into a new spreadsheet
  • In cell A1, type 'Gas', in B1 type 'Percentage of Earth's atmosphere'.
  • Continue until you have entered all the information shown above.
  If the information does not fit into some cells, then you may need to make some columns wider.
  To make column A wider:
  • Move the cursor to the right hand edge of the cell marked 'A'.
  • Hold down the left button on the mouse, and drag the mouse to the right until the column is the correct width.
  Repeat the same process with column B.

(b) Move the cursor to cell A1.

- Hold down the left button on the mouse, and drag the mouse to cell B4. All eight cells will now be highlighted.
- Click **Insert** on the toolbar, then **Chart**.
- Select **Pie** as your type of chart, and select the picture that matches the type of pie chart you would like. Now **Finish**.

The chart should look like the one below.

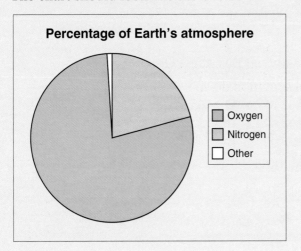

Avoid using the 3-D angled pie charts, or 3-D pie charts with slices pulled out. The data shown can be hard to interpret, or misleading.

Exercise 3K

1 In a local election, there were four candidates, Mr Angle, Mrs Chart, Mr Sector and Ms Pie. The table shows how many votes each candidate received in the election.

Candidate	Angle	Chart	Sector	Pie
Votes	3782	4187	3521	6225

Draw a pie chart to show this information.

2 The table below shows information about the goals scored by Lanchester Rovers last season.

Goalscorer	I. Missed	E. Scored	R.U. Sure	P.E. Nalty	Other
Number of goals	17	27	15	12	8

(a) How many goals did Lanchester Rovers score last season?

(b) Who was their top scorer?

(c) Draw a pie chart to show this information.

3 The table below shows the amount of money that is spent on Education, Health, Transport and Emergency services in the city of Suncastle.

Area of spending	Amount of money
Health	£57 000 000
Education	£62 000 000
Transport	£15 000 000
Emergency services	£34 000 000

(a) Construct a pie chart to display this information.

(b) In question 3 of Exercise 3F, this information was used to draw a pictogram. Which type of presentation do you prefer? Explain your answer.

4 A survey was conducted of the number of toilets found in each house in Hexham.
Some of the percentages are shown in the table.

Number of toilets	Percentage
1	47
2	
3	19
more than 3	6

(a) What percentage of houses had exactly two toilets?

(b) Construct a pie chart to show this information.

5 On a long car journey, Karen kept a record of the colours of all the cars that she had seen travelling in the opposite direction. The spreadsheet shows a record of her results.

	A	B
1	COLOUR	Frequency
2	Red	2396
3	Blue	1843
4	White	789
5	Black	558
6	Silver	387
7	Other	1256

(a) Copy this information onto a spreadsheet of your own.

(b) Use your spreadsheet to draw a pie chart for this information.

6 Use a spreadsheet to generate a pie chart for the data in question 2.

3.12 Interpreting pie charts

■ **You can calculate quantities represented by pie charts if you know the total quantity and the angles for each sector.**

Example 15

This pie chart shows the breakdown of 90 guests at a hotel.

(a) How many men were at the hotel?

(b) How many children were at the hotel?

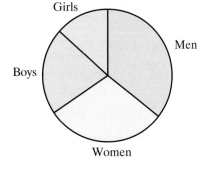

The whole pie chart represents 90 guests.

$360 \div 90 = 4°$

So each person is represented by 4°.

(a) The angle measures 128°.

$128 \div 4 = 32$ men.

(b) The combined angle for boys and girls measures 124°.

$124 \div 4 = 31$ children.

Sometimes you are told how many individuals are represented by one sector.

Example 16

This pie chart shows the colour of eyes of pupils in a school. 490 of the pupils have brown eyes.

(a) What is the size of the angle representing brown eyes?

(b) How many pupils are represented by each degree?

(c) How many pupils were in the school altogether?

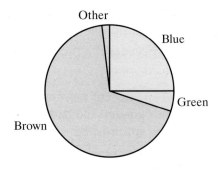

(a) With a protractor, the angle is 245°.

(b) $490 \div 245 = 2$
Each degree represents 2 pupils.

(c) $2 \times 360 = 720$ pupils.

Exercise 3L

1   72 dog owners were asked to name their preferred brand of dog food.
     The results are shown in the pie chart.

     (a) What angle will represent Barkie on the pie chart?

     (b) How many dog owners preferred Woofo?

     (c) How many dog owners preferred Ruff?

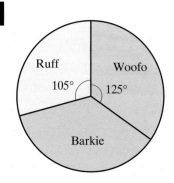

2 The pie chart below shows the 240 g of ingredients used to make a cake.

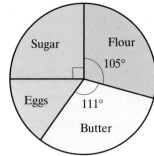

Ingredients	Weight
Flour	
Butter	
Eggs	
Sugar	

(a) Use the pie chart to complete the table.

(b) Use your answer to construct a bar chart for the same information.

3 The pie chart below shows the way Edward spent the last 24 hours.

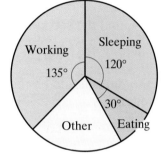

(a) How many degrees represents each hour in the day?

(b) What angle represents 'Other activities' in the pie chart?

(c) How many hours did Edward sleep yesterday?

(d) How many hours did Edward spend working yesterday?

4 Some children were asked the colour of their bedroom carpet.

Colour	Frequency	Angle
Red	8	72°
Blue	14	
Green	5	
Multi-coloured		36°
Other		81°
Total		

Copy and complete the table, and draw a pie chart to show this information.

3.13 Comparative pie charts

A pie chart shows *proportions*, but not actual frequencies.

Often, two sets of data have different total frequencies. If you draw two pie charts the same size, it can be misleading.

You can avoid this problem by making the pie charts different sizes. To make this completely fair, the areas of the two circles should be in the same ratio as the two frequencies.

■ Comparative pie charts can be used to compare two sets of data of different sizes. The areas of the two circles should be in the same ratio as the two frequencies.

Calculating the radius of a scaled pie chart

The area of a circle is calculated by using πr^2.

Call the radius of the first circle r_1, and the radius of the second circle r_2.

The ratio of the areas of the two circles will be:

$$\pi r_1^2 : \pi r_2^2$$
$$\text{or} \quad r_1^2 : r_2^2 \quad \text{(divide through by } \pi\text{)}$$

If you call the first frequency F_1, and the second frequency F_2, then

$r_1^2 : r_2^2 = F_1 : F_2$, which can be written as $r_1 : r_2 = \sqrt{F_1} : \sqrt{F_2}$

This means that the ratio of the radii should be the same as the ratio of the square roots of the frequencies.

Example 17

Residents of Appleton and Orangeford were asked how many television sets they had in their household.

The results are shown in the tables below.

Appleton

Number of TVs	Frequency
0	89
1	420
2	365
3	72
more than 3	18
Total	964

Orangeford

Number of TVs	Frequency
0	157
1	845
2	403
3	88
more than 3	30
Total	1523

Draw pie charts to show this information.

The ratio of the radii will be $\sqrt{964} : \sqrt{1523}$, or $31.05 : 39.03$.

It would be sensible to use a radius of 3.1 cm for the first pie chart, and a radius of 3.9 cm for the second pie chart.

Calculate the angles in the usual way; e.g. for Appleton:

0 TVs $\quad \dfrac{89}{964} \times 360 = 33.2° \ldots$

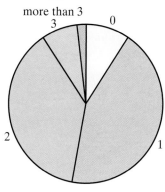

Appleton

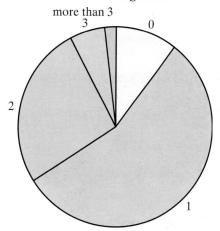

Orangeford

Exercise 3M

1   At Heidi High School, pupils can choose to study either French or Spanish. The tables below give information about their GCSE results.

French

GCSE result	Frequency
A or A*	12
B or C	36
D or E	24
F or G	9
Total	81

Spanish

GCSE result	Frequency
A or A*	8
B or C	29
D or E	9
F or G	3
Total	49

(a) Which was the most popular language at Heidi High school?
(b) What grades was a pupil most likely to score in a language GCSE at Heidi High school?
(c) Draw two scaled pie charts to show this information.

2   A survey was conducted to find the uses of land in two counties, Pinkshire and Perkshire. The results are shown in the table below.

Pinkshire

Use of land	Land area (acres)
Agriculture	89
Urban	420
Woodland	365
Water	72
Total	946

Perkshire

Use of land	Land area (acres)
Agriculture	157
Urban	845
Woodland	403
Water	88
Total	1493

(a) What is the most common use of land in Pinkshire?

(b) What is the most common use of land in Perkshire?

(c) Draw two scaled pie charts to show this information.

3 A survey was conducted to find out Europe's most popular classical composer.
The table shows the results of the survey in France and Germany.

Germany

Composer	Frequency
Beethoven	82
Mozart	486
Handel	136
Saint-Saëns	0
Wagner	48
Other	32

France

Composer	Frequency
Beethoven	255
Mozart	321
Handel	189
Saint-Saëns	287
Wagner	36
Other	33

Draw two scaled pie charts to show this information.

4

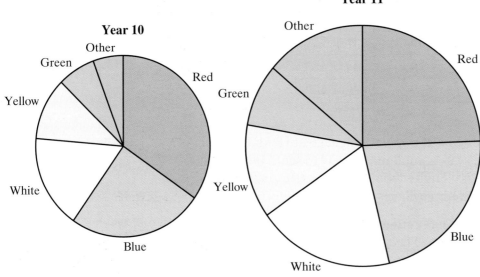

The pie charts show the favourite colours of two year groups in a school.
There are 180 in year 10.

(a) The radii of the two circles are 3 cm and 4 cm. How many students are there in Year 11?

(b) How many students liked the favourite colour in Year 10?

(c) How many students liked green in Year 11?

(d) Complete the table below for Year 10.

Colour	Frequency
Red	
Blue	
White	
Yellow	
Green	
Other	
Total	180

(e) How many students in Years 10 and 11 liked the colour blue?

(f) Use this information to draw a multiple bar chart to compare the favourite colours of the two year groups.

5 James and Alex both asked the children in each of their classes how they had travelled to school that morning.
The results are shown in the two tables below.

James' class

Mode of travel	Frequency
Bus	3
Train	2
Car	5
Bicycle	2
Walk	13
Total	25

Alex's class

Mode of travel	Frequency
Bus	3
Train	6
Car	11
Bicycle	1
Walk	15
Total	36

They draw pie charts to show their data. They are shown below.

James' class

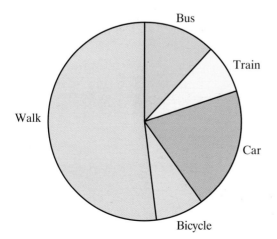

Alex's class

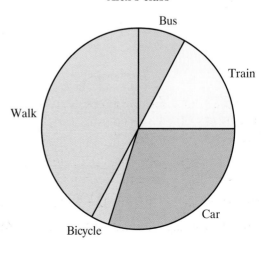

(a) These pie charts are misleading. At a glance it looks as if there were more children in James' class who walked to school. Explain why this is not the case.

(b) Using the frequency tables above, construct two comparative pie charts to show these data more effectively.

3.14 Stem and leaf diagrams

Here are the ages of 36 guests at a christening.

3	7	9	18	21	22	25	25	25	25	26	27
27	28	29	30	31	33	34	34	36	36	36	38
39	41	43	43	44	46	48	52	53	60	65	68

This information can be shown in a grouped frequency table, a bar chart or a pie chart.

Group	Frequency
0–9	3
10–19	1
20–29	11
30–39	10
40–49	6
50–59	2
60–69	3

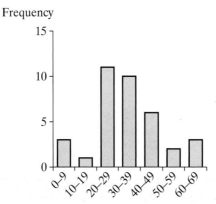

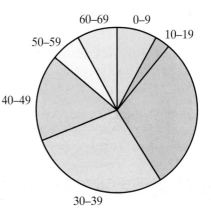

In each of these cases some of the information has been lost. For example, the frequency table shows you that there are 11 guests between the ages of 20 and 29, but you cannot tell the exact age of each guest.

■ **A stem and leaf diagram allows you to show the distribution in the same way as a bar chart but retains the detail of the data.**

Here is a stem and leaf diagram to show the ages of guests at the christening.

When drawing a stem and leaf diagram, put the leaves in numerical order.

The stems   The leaves   You can still see that 20–29 is the group with the highest frequency

```
0 | 3  7  9
1 | 8
2 | 1  2  5  5  5  5  6  7  7  8  9
3 | 0  1  3  4  4  6  6  6  8  9
4 | 1  3  3  4  6  8
5 | 2  3
6 | 0  5  8
```

Key
3 | 4 = 34 years old

You can tell that the oldest guest was exactly 68 years old.

Each combination of stem and leaf represents one of the guests. Add the leaf to its stem to find out the age of the guest. The stem gives the first digit and the leaf gives the second digit.

■ **A stem and leaf diagram must have a key that shows how the stem and leaf are combined together.**

Example 18

Here are the percentages scored by 30 pupils in a test.

8	12	17	27	35	36	38	42	42	44
45	45	45	47	49	53	57	59	62	68
73	77	78	80	83	84	89	92	97	98

Using a step of 10 for the stems, draw a stem and leaf diagram for this information.

The stems will be 0, 1, 2, 3, 4, 5, 6, 7, 8 and 9.
For example, 53 is represented by a stem of 5 and a leaf of 3.

```
0 | 8
1 | 2  7
2 | 7
3 | 5  6  8
4 | 2  2  4  5  5  5  7  9
5 | 3  7  9
6 | 2  8
7 | 3  7  8
8 | 0  3  4  9
9 | 2  7  8
```

Key
3 | 5 = 35%

Hint:
Remember the key!

You may be asked to find medians and quartiles. They can be found from stem and leaf diagrams.

Exercise 3N

1 Thirty members of a fitness club were asked how many 'sit-ups' they could do in a minute. The results are shown below.

12	15	16	23	26	26	27	28	29	29
32	33	33	33	35	37	38	39	40	40
41	42	45	48	53	59	68	72	75	239

   (a) Explain why the club manager decided to ignore the final result of 239.

   (b) Draw a stem and leaf diagram to show these data. Use steps of 10. Leave out the final result of 239.

   (c) Which number of sit-ups was the most common?

2 A shop manager records details of the customers during the first half hour that his shop is open.
   The stem and leaf diagram shows the ages of the customers.

```
0 | 1  2  4
0 | 5  6  6  8  9
1 | 1  1  2  2  2  3  4  4
1 | 5  5  6  7  8  8
2 | 2  3  3  4
2 | 5  5  8
3 | 2
```

Key
1 | 8 = 18 years old

(a) How many customers visited the shop in this time?

(b) What was the age of the oldest customer?

(c) What was the most common age of customer?

(d) How many customers were 6 years old?

(e) Draw a frequency table from these data.

(f) Redraw the stem and leaf diagram with steps of 10 between the stems.

(g) Draw a pie chart to show the same data.

(h) What data have been lost when constructing this pie chart?

3   Rebecca has a job that involves driving long distances each day. Here are the distances (to the nearest mile) that she travelled on each work day in January.

136	202	145	154	159	180	193	176	162	168	192
142	162	151	162	153	169	169	178	154	182	177

(a) Draw a stem and leaf diagram for this information.

(b) What other type of diagram could be used to show these data?

(c) What advantage does the stem and leaf diagram have over the type of diagram you have suggested for part (b)?

4   Thirty pupils were asked how many CDs they had in their collection.
The results are shown below.

23	2	18	14	7	4	25	21	18	15
32	26	31	6	17	6	18	19	22	23
31	21	12	1	0	8	14	15	18	26

Draw a stem and leaf diagram to show this information.

5   Carol asked 60 of her friends to name a whole number between 100 and 200.
Their answers are shown below.

100	107	134	140	152	153	152	124	148	132
162	163	173	104	122	145	144	145	147	105
156	103	102	135	142	155	156	114	123	134
172	178	171	172	175	109	127	143	146	109
151	154	106	106	149	148	155	157	112	129
171	179	170	177	103	128	144	147	149	102

Draw a stem and leaf diagram to show this information.

Revision exercise 3

1   An estate agent surveyed the number of bedrooms in each
    house on three streets of Frimmerton. The results of the survey
    are shown in the charts below.

Squib Street

Number of bedrooms	Frequency
1	3
2	17
3	22
4	11
5	7
Total	60

Crumple Street

Frequency

Number of bedrooms

Round Street

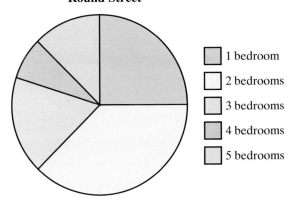

- 1 bedroom
- 2 bedrooms
- 3 bedrooms
- 4 bedrooms
- 5 bedrooms

(a) What is the most common number of bedrooms in Round
    street?

(b) How many two-bedroom houses are there in Crumple
    street?

(c) How many houses are on Squib Street?

(d) How many houses are there in total on Crumple Street?

(e) There are 48 houses on Round Street. How many of them
    have

    (i) 1 bedroom,   (ii) 2 bedrooms?

(f) Which street has the most four-bedroom houses?

(g) The estate agent said that 'In every street, there are more
    four-bedroom houses than there are five-bedroom houses.'
    Is he correct? Explain your answer.

(h) If you choose a house at random, in which street are you
    most likely to choose a house with exactly two bedrooms?

(i) Which type of chart do you think shows these data most
clearly?
Explain your answer.

(j) Show the data for Squib Street using

(i) a bar chart,
(ii) a pie chart,
(iii) a pictogram.

2   60 customers in a restaurant were
asked which part of their meal
they had enjoyed the most.
Their responses are shown in the
frequency table.

Town	Frequency
Starter	15
Main course	28
Dessert	10
Drinks	7

(a) Draw a pictogram to display
this information.

(b) Draw a pie chart to show
this information.

3   One Saturday, Adrian recorded the ages of the first 40
customers at a supermarket. The ages are shown below.

25	8	36	29	12	17	33	28	22	36
55	21	27	33	37	48	42	3	35	44
16	22	29	31	36	56	41	24	28	33
46	56	38	25	41	38	11	7	17	26

(a) Draw up a tally chart, using the class intervals 0–9, 10–19,
20–29 … etc.

(b) Why can you not use the intervals 0–10, 10–20, 20–30 … for
this tally chart?

(c) Construct a stem and leaf diagram to show these data.

(d) What information was lost when constructing the tally
chart, but could still be seen on the stem and leaf
diagram?

4   The number of boys and girls in each year at Finbow High
School is shown in a two-way table.

	Year 7	Year 8	Year 9	Year 10	Year 11	Total
Boys	72		71	66		320
Girls	63	75		55		286
Total		122	101			

(a) Copy and complete the two-way table.

(b) Draw a multiple bar chart to show the number of boys and
girls in each year of Finbow High School.

5 Debbie asked 40 adults to choose their favourite polygon, and then classified their answers by the number of sides. The results are shown on the section of a spreadsheet below.

	A	B
1	Type of shape	Frequency
2	Triangle	9
3	Quadrilateral	13
4	Hexagon	8
5	Octagon	6
6	Other	4
7	Total	40

(a) Copy this information onto a spreadsheet of your own.

(b) Use your spreadsheet to draw a bar chart to show this information.

(c) Use your spreadsheet to draw a pie chart to show this information.

Summary of key points

1 A tally chart, or frequency table, can be used to process raw data, making it easier to spot patterns.

2 When the data are widely spread, you should group the data into classes.

3 Choose an appropriate number and width of class intervals in order to make any patterns and trends more obvious. Between 5 and 10 class intervals will often be appropriate.

4 When there is not an even spread of data across the range use class intervals of varying width.

5 When the extreme values of the data are not known, leave the first and/or last intervals open.

6 A two-way table enables you to show two variables at the same time.

7 Organising data into a two-way table can help you to calculate missing information.

8 A database is a collection of information. Computers can store huge databases that can be easily selected, sorted and ordered at the touch of a button.

9 A summary table shows data that have been sorted and summarised, and is easier to interpret than the original raw data.

10   Diagrams and charts are used to represent data in a clear and simple way, but do not always show the exact information.

11   A pictogram uses symbols to represent a certain number of items.

12   Bar charts are a simple way to show trends for discrete data. The bars may be horizontal or vertical.

13   A vertical line graph can be used to display discrete data. The format is very similar to a bar chart, with a series of vertical lines evenly spaced.

14   Multiple bar charts have more than one bar for each class.

15   A compound bar chart has single bars split into separate sections for each category. You can compare groups, and you can see the distribution of the data as a whole.

16   A pie chart is a good way of displaying data when you want to show how something is shared or divided.

17   When constructing pie charts for real data, you will often need to use rounded values for the size of the angles.

18   You can calculate quantities represented by pie charts if you know the total quantity and angles for each sector.

19   Comparative pie charts can be used to compare two sets of data of different sizes. The areas of the two circles should be in the same ratio as the two frequencies.

20   A stem and leaf diagram allows you to show the distribution in the same way as a bar chart but retains the detail of the data.

21   A stem and leaf diagram must have a key that shows how the stem and leaf are combined together.

4 Representing and processing continuous data

4.1 Frequency tables for continuous data

Thirty students timed how many seconds they could hold their breath to the nearest one hundredth of a second. Here are the raw data:

30.15	42.92	26.48	48.14	9.70	31.05	42.59	20.00
29.83	19.40	36.73	43.66	29.92	33.50	14.31	26.37
48.06	31.18	49.37	21.68	51.30	58.39	37.52	
7.34	36.89	27.84	39.18	16.50	39.42	12.31	

Shaun, Rita and Philip each tried to sort the data into a frequency table. They each wrote their class intervals differently.

Shaun

Time (seconds)
0 to 9
10 to 19
20 to 29
30 to 39
40 to 49
50 to 59

Rita

Time (seconds)
0–10
10–20
20–30
30–40
40–50
50–60

Philip

Time, t (seconds)
$0 < t \leqslant 10$
$10 < t \leqslant 20$
$20 < t \leqslant 30$
$30 < t \leqslant 40$
$40 < t \leqslant 50$
$50 < t$

Shaun found difficulty when recording the number 9.70. It did not fit in either '0 to 9' or '10 to 19'.

Shaun's table is unsuitable for continuous data because there are gaps between groups.

Rita found difficulty when recording the number 20.00. It fits into both '10–20' and '20–30'.

Rita's table is unsuitable because there is an overlap between groups.

Philip's class boundaries do not overlap, and have no gaps. The inequalities show that 20.0 falls in the group '$10 < t \leqslant 20$', but not in the group '$20 < t \leqslant 30$'.

Time, t (seconds)	Tally	Frequency
$0 < t \leqslant 10$	||	2
$10 < t \leqslant 20$	卌	5
$20 < t \leqslant 30$	卌		6
$30 < t \leqslant 40$	卌 ||||	9
$40 < t \leqslant 50$	卌		6
$50 < t$	||	2

Philip left the last class open, as it was difficult to estimate the longest time for which a student could hold his or her breath.

■ **When sorting continuous data into a frequency table, class intervals must be joined but not overlapping. Inequalities can be used to define the class boundaries.**

Varied class widths can be used when there is not an even spread of data across the range. Use narrower groups where the data are clustered. Use wider groups where data are spread out.

Exercise 4A

1 Gareth records the amount of rainfall that falls each day in Runsby. Here is the raw data for January (in cm).

5.6	4.3	2.1	0	0.8	5.2	3.3	2.8
2.2	1.6	0.4	1.9	3.2	4.2	1.0	3.0
3.6	2.4	1.8	0.4	0	0	3.2	3.5
2.7	1.2	2.1	1.1	5.7	5.2	3.1	

Gareth decides to sort his results using the following frequency table.

Rainfall, r (cm)	Tally	Frequency
$0 \leqslant r < 1$		
$1 \leqslant r < 2$		
$2 \leqslant r < 3$		
$3 \leqslant r < 4$		
$4 \leqslant r < 5$		
$5 \leqslant r < 6$		

(a) In which class will Gareth put a rainfall of 1 cm?

(b) Copy and complete the frequency table.

(c) Draw a pie chart to display Gareth's data.

2 Frank has a collection of 34 garden gnomes.
The weight in kilograms of each of the garden gnomes is shown
below (to 2 decimal places).

2.44 1.57 2.35 1.13 2.52 1.59 2.53 0.65 2.56
1.60 2.67 1.22 2.89 1.72 2.99 0.27 3.00 1.77
3.13 1.34 3.22 1.81 0.74 1.88 1.37 1.91 0.48
2.11 1.48 2.36 0.85 2.22 1.53 2.29

(a) What is the weight of the heaviest gnome?
(b) Frank begins to draw a frequency table:

Weight, w (kg)	Tally	Frequency
$0 \leqslant w < 0.5$		
$0.5 \leqslant w < 1$		

Copy and complete Frank's frequency table, with classes of
equal width.

3 Helen and Nigel decide to measure the capacity (in litres) of the
buckets belonging to children on a beach.
They decide to use the class intervals below.

Helen

Capacity (litres)
0 to 0.4
0.5 to 0.9
1.0 to 1.4
1.5 to 1.9
2.0 to 2.9

Nigel

Capacity (litres)
0–0.5
0.5–1.0
1.0–1.5
1.5–2.0
2.0–2.5

(a) One of the buckets had a capacity of 1.46 litres.
Explain why Helen would have trouble recording this using
her class intervals.
(b) One of the buckets had a capacity of exactly 1.5 litres.
Explain why Nigel would have trouble recording this using
his class intervals.
(c) Write the class intervals differently, so that these problems
would not occur.

4 Police at a checkpoint measured the speed of passing cars in
kilometres per hour. Here are the results for the first 30 cars.

52.8 60.3 70.4 32.3 53.8 71.6 63.2 73.9 42.3
78.2 64.4 63.8 54.1 56.4 52.7 48.6 65.3 66.2
65.9 43.9 38.5 57.7 41.3 59.3 47.8 68.7 62.9
75.7 65.3 76.8

Design and complete a frequency table to sort this information.

5 One event at a local fete was a 'throw the welly' competition. The frequency table below shows some information about the distances the wellies were thrown.

Distance thrown, d (metres)	Frequency
$0 < d \leqslant 5$	3
$5 < d \leqslant 10$	0
$10 < d \leqslant 15$	0
$15 < d \leqslant 20$	4
$20 < d \leqslant 25$	6
$25 < d \leqslant 30$	12
$30 < d \leqslant 35$	26
$35 < d \leqslant 40$	37
$40 < d \leqslant 45$	28
$45 < d \leqslant 50$	8
$50 < d \leqslant 55$	4
$55 < d \leqslant 60$	1
$60 < d \leqslant 65$	0
$65 < d \leqslant 70$	1

(a) Explain how this frequency table could have been better with unequal class intervals.

(b) By combining some intervals, draw out a more compact frequency table. Use no more than 7 (unequal) class intervals.

(c) Draw a pie chart to show this information.

4.2 Rounded data

■ When raw data are rounded before they are recorded, set class boundaries so that all possible values of a rounded number fit into the same class.

Example 1

Here are the times taken for 28 people to run 400 metres. The times are rounded to the nearest second.

54	58	69	82	70	51	55
63	68	77	53	66	58	72
52	56	71	76	63	81	75
83	72	68	55	61	68	67

Design a frequency table to record these continuous data.

The data do not show the actual times taken – the real values have been rounded. The time t shown as 70 seconds could have been anything in the range

$$69.5 \leqslant t < 70.5$$

You must take this into account when setting class boundaries.

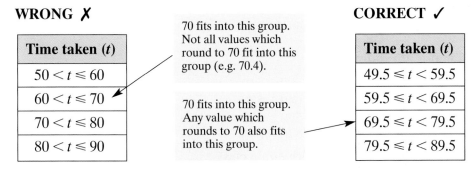

WRONG ✗

Time taken (t)
$50 < t \leqslant 60$
$60 < t \leqslant 70$
$70 < t \leqslant 80$
$80 < t \leqslant 90$

70 fits into this group. Not all values which round to 70 fit into this group (e.g. 70.4).

70 fits into this group. Any value which rounds to 70 also fits into this group.

CORRECT ✓

Time taken (t)
$49.5 \leqslant t < 59.5$
$59.5 \leqslant t < 69.5$
$69.5 \leqslant t < 79.5$
$79.5 \leqslant t < 89.5$

Time taken (t)	Frequency
$49.5 \leqslant t < 59.5$	9
$59.5 \leqslant t < 69.5$	9
$69.5 \leqslant t < 79.5$	7
$79.5 \leqslant t < 89.5$	3

Exercise 4B

1   Here are the weights of 30 boys rounded to the nearest kilogram.

60	62	51	53	42	52	50	53	48	55
58	59	63	49	52	54	35	53	44	54
46	57	46	67	58	56	48	48	37	41

 (a) What is the range of values that could be represented by the weight 48 kg?

 (b) Kathleen wanted to use class intervals of $35 \leqslant t < 40$, $40 \leqslant t < 45$, $45 \leqslant t < 50$, etc. Explain why this is wrong.

 (c) Choose class intervals of width 5 kg that would suit these rounded data.

 (d) Use your class intervals from (c) to create and complete a frequency table for these data.

2   A breakdown service claims that, 'We reach our customers in less than half an hour in more than 90% of cases.'
    Here are the times to the nearest minute of 21 call outs.

27	11	15	28	34	22	5	26	19	22	8
28	25	27	22	33	21	29	18	14	29	

(a) Design and complete a frequency table with equal class intervals for these data.

(b) Can the breakdown service justify their claim? Explain your answer.

3   While training for a long jump competition, Sarah completes 36 practice jumps.
The length of each jump is measured to the nearest ten centimetres.
The lengths of the 36 jumps are shown below.

630	520	740	660	540	520	640	590	640
580	590	720	660	490	510	720	640	570
530	610	660	620	580	430	540	580	610
650	750	530	480	570	530	610	640	670

(a) Design and complete a frequency table with equal class intervals of width 50 cm for these data.

(b) The winning jump at last year's competition was 717 cm. How many of Sarah's practice jumps would definitely have beaten last year's winner?

4   The Maths Social Society held a competition to test the strength of their members. The first event involved holding a heavy object for as long as possible.
The times were measured to the nearest tenth of a second.
Joff, Lee and Bernie decided to keep a frequency table to record the results, using the class intervals shown below.

Joff

Time (seconds)
0 to 9
10 to 19
20 to 29
30 to 39
40 to 49
50 to 59

Lee

Time (seconds)
$0 < t \leqslant 10$
$10 < t \leqslant 20$
$20 < t \leqslant 30$
$30 < t \leqslant 40$
$40 < t \leqslant 50$
$50 < t$

Bernie

Time, t (seconds)
$0 < t \leqslant 10.05$
$10.05 < t \leqslant 20.05$
$20.05 < t \leqslant 30.05$
$30.05 < t \leqslant 40.05$
$40.05 < t \leqslant 50.05$
$50.05 < t \leqslant 60.05$

(a) The first contestant held the object for 9.7 seconds. Explain why Joff would have trouble recording this in his table.

(b) The second contestant had his time rounded to 20.0 seconds. Lee did not know which class to record this in. Explain why.

(c) Bernie has more sensible class intervals for rounded data. Before the competition, it was not clear what the longest time would be.
Explain how Bernie could change her last class interval.

4.3 Frequency polygons

Once you have grouped and processed your data, you will usually want to show the information in a diagram or graph.

■ **A frequency polygon is used to show the shape of a continuous frequency distribution.**

A **frequency polygon** uses the mid-point of a class interval to represent all the data in that interval.

Plot the mid-point for each class interval against the frequency for that interval. Join the points with straight lines.

Example 2

Pupils were asked to say their twelve times table as quickly as possible.
The table shows the frequency distribution of the times taken.

Time (t seconds)	$0 < t \leqslant 10$	$10 < t \leqslant 20$	$20 < t \leqslant 30$	$30 < t \leqslant 40$	$40 < t \leqslant 50$	$50 < t \leqslant 60$
Frequency	1	2	8	12	6	3

Draw a frequency polygon to show this information.

Plot against the mid-point of the interval.

For more difficult examples the mid-point can be found by adding the two class boundaries together and dividing by two.
Here the mid-point is

$$\frac{40 + 50}{2} = 45$$

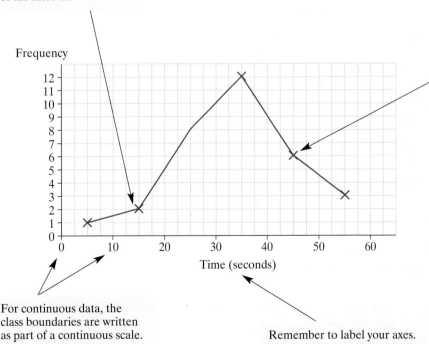

For continuous data, the class boundaries are written as part of a continuous scale.

Remember to label your axes.

For comparison, two similar frequency polygons can be drawn on the same graph.

4.4 Histograms with equal class intervals

Another diagram that can be used to represent a frequency distribution is a **histogram**.

■ **A histogram is made up of a series of bars or rectangles. The area of each rectangle represents the frequency of a class interval.**

When you have equal class intervals, a histogram looks like a bar chart where the bars are joined.

Here is a histogram to show the same data as the frequency polygon:

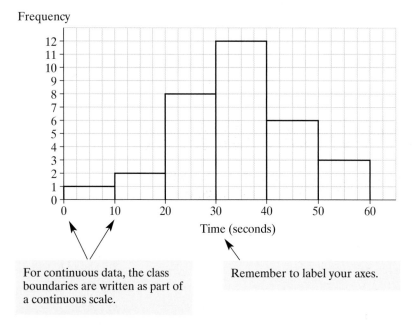

For continuous data, the class boundaries are written as part of a continuous scale.

Remember to label your axes.

A histogram is often drawn as a guide, so that a frequency polygon can be drawn over the top.

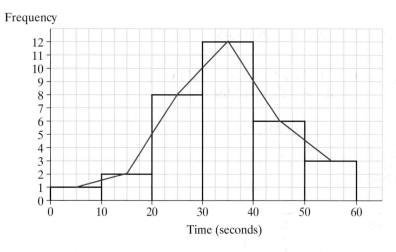

■ **For a histogram or frequency polygon, the class boundaries are written as part of a continuous scale.**

Many frequency polygons and histograms make the shape of a
normal distribution.

See section 9.4

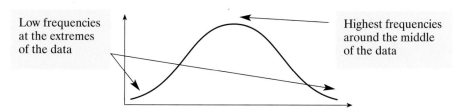

Low frequencies
at the extremes
of the data

Highest frequencies
around the middle
of the data

Often the shape of the distribution is **skewed** to one side.

Positive skew   Negative skew

See section 9.4

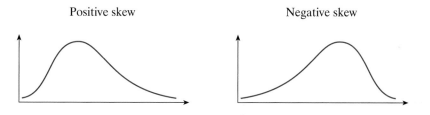

Exercise 4C

1   Last season, Gander United played 30 football matches away
    from home. The table gives information about the distances
    travelled to away matches.

Distance, m (miles)	Frequency
$0 < m \leqslant 20$	6
$20 < m \leqslant 40$	12
$40 < m \leqslant 60$	7
$60 < m \leqslant 80$	4
$80 < m \leqslant 100$	1

Draw a frequency polygon to display this information.

2   Amy decided to measure the height of each of her 50 cuddly toys.
    The table gives information about the heights of the toys.

Height (cm)	Frequency
up to but not including 10	3
10 up to but not including 20	12
20 up to but not including 30	19
30 up to but not including 40	10
40 up to but not including 50	6

Draw a frequency polygon to display this information.

3   Frequency

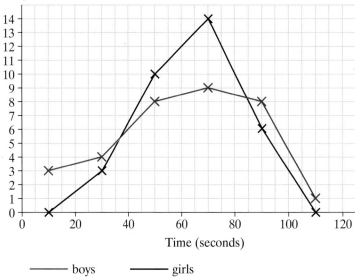

— boys   — girls

Students were asked to balance a wooden block on their heads for as long as possible. The frequency polygon gives information about the performance of boys and girls in this task.

(a) How many girls were able to balance the block between 20 and 40 seconds?

(b) How many boys let the block fall in less than or equal to 40 seconds?

(c) One pupil balanced the block longer than the other pupils. Was this a boy or a girl?

(d) Copy and complete the frequency table below.

Time, t (seconds)	Girls' frequency	Boys' frequency
$0 < t \le 20$		
$20 < t \le 40$		
$40 < t \le 60$		
$60 < t \le 80$		
$80 < t \le 100$		
$100 < t \le 120$		

(e) Were there more boys or girls in the experiment?

(f) Who were best at balancing, boys or girls? Explain your answer.

(g) Draw a histogram to represent the boys' data.

4 The heights of Year 7 and 8 pupils were measured.
The table shows some information about the results.

Height, h (centimetres)	Year 7 frequency	Year 8 frequency
$120 < h \leqslant 130$	1	0
$130 < h \leqslant 140$	5	3
$140 < h \leqslant 150$	18	12
$150 < h \leqslant 160$	20	22
$160 < h \leqslant 170$	8	19
$170 < h \leqslant 180$	2	6

(a) Draw a histogram to show Year 7 heights.
(b) Draw a histogram to show Year 8 heights.
(c) Draw a frequency polygon to compare the heights of pupils in Years 7 and 8.

5 30 pupils were asked to time their journey to school to the nearest minute. The results are shown below.

```
 6   18   29   55    7   34   28   56   33    4
 2   41   33   23    7   43   26   53   44   41
32   46   16   17    3   26   17   47   22   17
```

(a) Using class intervals of equal width, design and complete a frequency table to sort these data. (Remember: the data are rounded.)
(b) Draw a histogram to display your results.

4.5 Histograms with unequal class intervals

When continuous data are sorted into unequal class intervals, it would be misleading to draw a histogram with frequencies as heights.

Consider these data:

A histogram using frequencies would look like this:

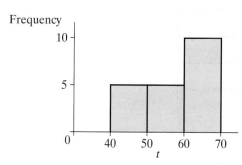

Class	Frequency
$40 < t \leqslant 50$	5
$50 < t \leqslant 60$	5
$60 < t \leqslant 70$	10

If the first two classes are combined you get:

Class	Frequency
$40 < t \leqslant 60$	10
$60 < t \leqslant 70$	10

The histogram would now look like this:

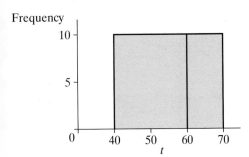

It looks as if there are twice as many between 40 and 60 as there are between 60 and 70. This is misleading.

Example 3

The frequency table shows the time taken for each of 400 people to climb up 8 flights of stairs.

Time, t (seconds)	Frequency
$40 < t \leqslant 60$	100
$60 < t \leqslant 70$	60
$70 < t \leqslant 80$	90
$80 < t \leqslant 85$	70
$85 < t \leqslant 90$	60
$90 < t \leqslant 120$	90

The class interval $40 < t \leqslant 60$ appears to be the most popular, as it has the highest frequency.

Although this is only a frequency of 70, it is spread across a time of only 5 seconds.

You need to take into account the width of each class interval. Add two columns to your table and calculate the **frequency density** as shown.

■ **In a histogram, frequency density = frequency ÷ class width**

Time, t (seconds)	Frequency	Class width	Frequency density
$40 < t \leqslant 60$	100	20	5
$60 < t \leqslant 70$	60	10	6
$70 < t \leqslant 80$	90	10	9
$80 < t \leqslant 85$	70	5	14
$85 < t \leqslant 90$	60	5	12
$90 < t \leqslant 120$	90	30	3

$60 - 40 = 20$

$90 \div 10 = 9$

You can use a histogram to display these data.

■ **In a histogram, use frequency density for the heights of the rectangles. Make sure the rectangle spreads across the entire class interval.**

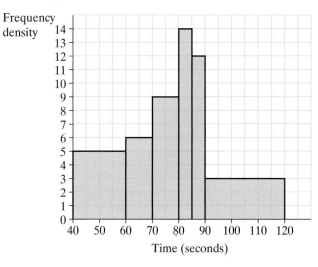

Frequency density

Time (seconds)

Frequency density replaces frequency on the vertical axis.

The area of each rectangle is the same as the frequency for that class interval.

It is not necessary to write values for frequency density on the vertical axis – you can use a key. The key tells you what a particular area represents. In this case the key would be:

□ = 5 people

Using the definition of frequency density given above,

■ **The areas of the rectangles are equal to the frequencies they represent.**

Example 4

Tony likes to jog in his spare time. He records how many miles he runs each week (to the nearest mile).

Here are the distances for the last six months.

$$\begin{array}{ccccccccccc} 2 & 27 & 12 & 36 & 12 & 14 & 13 & 31 & 5 & 17 \\ 33 & 12 & 18 & 19 & 18 & 25 & 4 & 14 & 17 & 19 \\ 28 & 17 & 14 & 19 & 25 & 3 \end{array}$$

Draw a histogram to display these data.

First you need to sort the data into a frequency table.

- There are not many distances below 10 miles or above 20 miles (wide groups).

- There are a lot of distances in the range 12 to 19 miles (narrow groups).

- Remember that the continuous data have been rounded.

Distance, d (miles)	Tally	Frequency
$0.5 < d \leqslant 10.5$	||||	4
$10.5 < d \leqslant 15.5$	ⅠⅡⅡ ||	7
$15.5 < d \leqslant 20.5$	ⅠⅡⅡ |||	8
$20.5 < d \leqslant 30.5$	||||	4
$30.5 < d \leqslant 40.5$	|||	3

Now use the frequency and class width to find the frequency density.

Distance, d (miles)	Frequency	Class width	Frequency density
$0.5 < d \leqslant 10.5$	4	10	$4 \div 10 = 0.4$
$10.5 < d \leqslant 15.5$	7	5	$7 \div 5 = 1.4$
$15.5 < d \leqslant 20.5$	8	5	$8 \div 5 = 1.6$
$20.5 < d \leqslant 30.5$	4	10	$4 \div 10 = 0.4$
$30.5 < d \leqslant 40.5$	3	10	$3 \div 10 = 0.3$

You now have the information you need to draw the histogram.

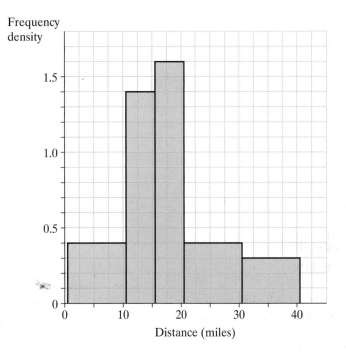

Frequency density

Distance (miles)

Note: When drawing the bars, use the class bounds.

Exercise 4D

1 The lengths of 62 songs by Median and the Meanies are represented in the frequency table.

Length of song, L (seconds)	Frequency
$0 < L \leqslant 100$	5
$100 < L \leqslant 180$	8
$180 < L \leqslant 210$	12
$210 < L \leqslant 240$	15
$240 < L \leqslant 300$	12
$300 < L \leqslant 500$	10

Draw a histogram to display these data.

2 The exact ages (in years) of guests at a fancy dress party were recorded. Some information about the guests' ages is shown in the frequency table.

Age, A (years)	Frequency
$15 < A \leqslant 20$	5
$20 < A \leqslant 23$	15
$23 < A \leqslant 25$	20
$25 < A \leqslant 30$	20
$30 < A \leqslant 40$	10

(a) Draw a histogram to display these data.

(b) Use your histogram as a guide to draw a frequency polygon for these data.

3 Alice measures the heights of 30 students in her class to the nearest centimetre.
The raw data are shown below.

```
141  156  168  179  143  158  155  154  153  149
162  178  164  154  153  166  153  168  143  159
157  156  157  165  178  175  152  154  149  175
```

(a) Design and complete a suitable frequency table with unequal class intervals to display these data.

(b) Draw a histogram to represent these data.

4 Justin conducted an experiment to see how far 33 snails would travel in a period of 10 minutes. The results are shown in the frequency table below.

Distance travelled, d (cm)	Frequency
$0 < d \leqslant 5$	3
$5 < d \leqslant 7$	5
$7 < d \leqslant 8$	4
$8 < d \leqslant 9$	6
$9 < d \leqslant 10$	3
$10 < d \leqslant 15$	6
$15 < d \leqslant 25$	6

Construct a histogram to display this information.

5 Measure the handspan of members of your class to the nearest millimetre.

(a) Record these data in a frequency table with unequal class intervals.

(b) Display these data using a histogram.

4.6 Reading from a histogram with unequal class intervals

The formula that we have used to calculate frequency density can be rearranged.

■ **Frequency = frequency density × class width**

Example 5

Use the histogram to complete a frequency table.

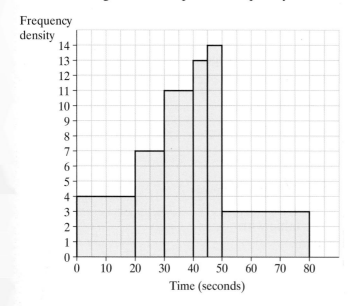

| First use the horizontal axis to determine the class intervals. | Then use the class boundaries to write down the class width. | Now read the frequency density from the vertical axis. | Frequency = frequency density × class width |

Time, t (seconds)	Class width	Frequency density	Frequency
$0 < t \leqslant 20$	20	4	80
$20 < t \leqslant 30$	10	7	70
$30 < t \leqslant 40$	10	11	110
$40 < t \leqslant 45$	5	13	65
$45 < t \leqslant 50$	5	14	70
$50 < t \leqslant 80$	30	3	90

Sometimes there are no values for the frequency density on the vertical axis. If you know at least one frequency, you can calculate the other frequencies.

Example 6

Complete the frequency table from the histogram.

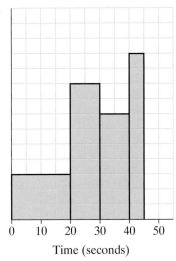

Time, t (seconds)	Frequency
$0 < t \leqslant 20$	36
$20 < t \leqslant 30$	
$30 < t \leqslant 40$	
$40 < t \leqslant 45$	

The interval $0 < t \leqslant 20$ has a frequency of 36.

This is represented on the histogram by 12 squares.

$36 \div 12 = 3$, so each square on the histogram counts for a frequency of 3.

Time, t (seconds)	Frequency
$0 < t \leqslant 20$	36
$20 < t \leqslant 30$	18 squares $\times$ 3 = 54
$30 < t \leqslant 40$	14 squares $\times$ 3 = 42
$40 < t \leqslant 45$	11 squares $\times$ 3 = 33

Exercise 4E

1   The histogram gives information about the distances (in metres) thrown in a javelin competition.

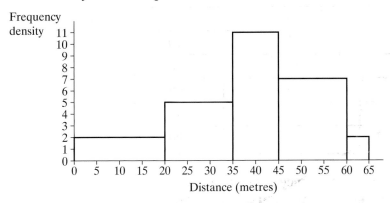

Design and complete a frequency table for the data.

2   A group of children time how long (in seconds) they can juggle a
    ball. The histogram gives information about their times.

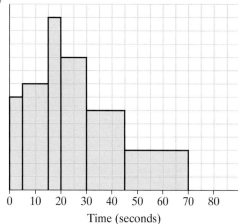

Time, t (seconds)	Frequency
$0 < t \leqslant 5$	56
$5 < t \leqslant 15$	
$15 < t \leqslant 20$	
$20 < t \leqslant 30$	
$30 < t \leqslant 45$	
$45 < t \leqslant 70$	
$70 < t \leqslant 80$	24

(a) Copy and complete the table.

(b) Copy and complete the histogram for the final class.

4.7 Stem and leaf diagrams for continuous data

■   **A stem and leaf diagram allows you to show the distribution in
    the same way as a frequency diagram, retaining the detail of the
    data.**

As for discrete data, the stems move up in equal steps. Each item of
data is represented by a combination of a stem with one of its leaves.

Example 7

The depth of rainfall in Ramsberry was measured each day of
December.
Here are the raw data (in cm).

2.3	1.6	0.7	1.2	3.1	4.5	1.0	3.2
3.5	2.3	1.7	0.4	0.0	0.0	3.0	0.0
2.7	1.2	2.1	1.1	2.8	5.2	1.1	5.7
2.1	0.0	0.8	5.6	4.3	5.2	3.3	

Using a step of 1 cm, draw a stem and leaf diagram for this information.

The stems will be 0, 1, 2, 3, 4 and 5.
For example, 2.3 is represented by a stem of 2 and a leaf of 3.
The data provided are not in order. First, you may find it useful to draw an unordered stem and leaf diagram.

```
0 | 7  4  0  0  0  0  8
1 | 6  2  0  7  2  1  1
2 | 3  3  7  1  8  1
3 | 1  2  5  0  3
4 | 5  3
5 | 2  7  6  2
```

Notice that 3 is represented by 3 | 0, i.e. 3.0

Key
1 | 6 = 1.6 cm of rainfall

From this unordered diagram, it is now easy to draw a fully ordered stem and leaf diagram.

```
0 | 0  0  0  0  4  7  8
1 | 0  1  1  2  2  6  7
2 | 1  1  3  3  7  8
3 | 0  1  2  3  5
4 | 3  5
5 | 2  2  6  7
```

Key
1 | 6 = 1.6 cm of rainfall

Exercise 4F

1 30 people were asked to close their eyes and raise their hand when they thought that 20 seconds had passed. The times are recorded (in seconds).

17.3	17.5	17.8	18.2	18.3	18.5	18.6	18.8	18.9	19.3
19.4	19.6	19.7	19.8	19.9	20.2	20.2	20.3	20.5	20.6
20.7	20.8	20.9	21.3	21.5	21.8	21.9	22.0	22.3	22.8

(a) Which estimate was the nearest?
(b) Draw a stem and leaf diagram to show these data. Use steps of 1 second.

2 As part of a survey, the heights of men visiting a clothes shop over a period of one hour were recorded. The stem and leaf diagram shows the heights of the customers (in metres).

```
1.5 | 1  6  8
1.6 | 3  4  6  8  9
1.7 | 1  2  2  4  6  7  7  8  9
1.8 | 2  3  3  4  5  5  7
1.9 | 1
```

Key
1.5 | 6 = 1.56 metres

(a) How many customers visited the shop in this time?
(b) What was the height of the tallest customer?
(c) How many customers were at least 170 cm tall?
(d) Draw a frequency table from these data.
(e) Construct a pie chart to display these data.

3 Catherine measures the diameters of small trees in a nursery.
When the trees have a diameter of over 6 cm, they are planted
in the forest.
Here are the diameters of 50 trees in Catherine's nursery.

4.5	2.1	2.5	3.4	5.6	5.7	4.9	2.2	4.7	2.3
5.9	6.0	6.4	5.0	2.6	2.3	2.5	2.7	5.8	5.0
4.2	2.8	3.0	3.4	3.3	3.7	4.6	3.3	3.1	3.5
4.3	3.7	3.7	3.9	3.9	4.3	4.6	4.8	4.9	5.0
5.3	5.3	5.6	3.7	3.3	0.2	5.8	5.9	6.2	6.3

(a) Draw an unordered stem and leaf diagram for these data.

(b) Use part (a) to help you draw an ordered stem and leaf
diagram for these data.

(c) How many trees were ready to be planted?

4

2	0	1	3	6	8		
3	1	2	2	2	6	9	
4	0	1	3	7	7		
5	1	2	3	4			
6	3						

Key

2 | 3   2.3 seconds to
answer the question

This stem and leaf diagram shows the lengths of time it took
people to answer a general knowledge question.

(a) What is the most common amount of time taken to answer
the question (to the nearest tenth of a second)?

(b) In total, how many people answered the question?

(c) What is the range of times taken to answer the question?

(d) What advantage does this stem and leaf diagram have over
a frequency table for the same data?

5 In a sprint race, the *reaction time* is the time it takes a sprinter to
start to move after the gun has fired to start the race.
The figures below show the reaction times, in seconds, of the
32 sprinters who took part in the first round of a competition (to
2 decimal places).

0.19	0.22	0.48	0.17	0.27	0.25	0.28	0.33
0.18	0.27	0.28	0.27	0.19	0.36	0.24	0.35
0.42	0.33	0.27	0.18	0.16	0.28	0.22	0.27
0.38	0.32	0.23	0.33	0.17	0.49	0.44	0.24

Draw an ordered stem and leaf diagram to show these data. Use
stems of 0.1, 0.2, 0.3, 0.4 and 0.5.

4.8 Population pyramids

The population of a country or area can be classified in many
different ways.

■ **A population pyramid allows you to compare aspects of a
population, usually by gender.**

A **population pyramid** looks similar to two bar charts back to back.

Example 8

The table shows the percentage of the population of a country in each age group.

Age, a (years)	$0 < a \leqslant 10$	$10 < a \leqslant 20$	$20 < a \leqslant 30$	$30 < a \leqslant 40$	$40 < a \leqslant 50$
Male %	13.8	11.9	16.5	19.4	13.7
Female %	12.4	13.6	15.2	17.4	12.8

Age, a (years)	$50 < a \leqslant 60$	$60 < a \leqslant 70$	$70 < a \leqslant 80$	$a > 80$
Male %	9.5	8.2	5.6	1.3
Female %	8.6	7.9	8.4	3.7

Display this information on a population pyramid.

To avoid giving a misleading impression, it is important that the class intervals are of equal size.

From this population pyramid it is possible to see that:

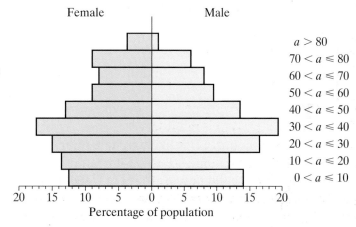

- there is a higher concentration of the population between the ages of 30 and 40 years,
- higher proportions of females live to be over 70 years old,
- there has been a greater proportion of boys than girls born over the last 10 years.

Population pyramids can show actual frequencies instead of percentages. You can also use them to compare other aspects of the population, e.g. height.

Exercise 4G

1   The table shows the percentage of the population of a country in each age group.

Age, a (years)	$0 < a \leqslant 10$	$10 < a \leqslant 20$	$20 < a \leqslant 30$	$30 < a \leqslant 40$	$40 < a \leqslant 50$
Male %	24	20	17	14	11
Female %	21	17	15	12	10

Age, a (years)	$50 < a \leqslant 60$	$60 < a \leqslant 70$	$70 < a \leqslant 80$	$a > 80$
Male %	8	3	2	1
Female %	10	8	5	2

Display this information on a population pyramid.

2

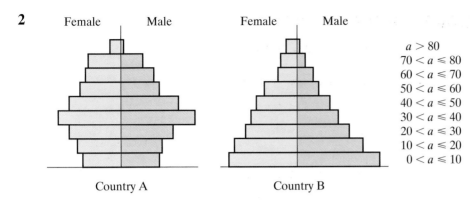

The population pyramids give information about the ages of the populations in two countries.

(a) Which country has had a high birth rate in recent years?

(b) In both countries, are 'long livers' more likely to be male or female?

(c) If a person was chosen at random from country A, what age are they most likely to be?

(d) In which country are people more likely to live beyond 50 years old?

(e) Explain why country B is likely to be a poor country.

3 The table shows percentages of over-18s classified by height.

Height, h (cm)	$130 < h \leqslant 135$	$135 < h \leqslant 140$	$140 < h \leqslant 145$	$145 < h \leqslant 150$	$150 < h \leqslant 155$	$155 < h \leqslant 160$
Male %	0.5	2	3	8	10	13
Female %	2	6	7	11	13	15

Height, h (cm)	$160 < h \leqslant 165$	$165 < h \leqslant 170$	$170 < h \leqslant 175$	$175 < h \leqslant 180$	$180 < h \leqslant 185$	$h > 185$
Male %	20	18	15	6	3	1.5
Female %	15	14	11	4	1.5	0.5

Display this information on a population pyramid.

4 The table shows the percentage of the population of a country in each age group.

Age, a (years)	$0 < a \leqslant 10$	$10 < a \leqslant 20$	$20 < a \leqslant 30$	$30 < a \leqslant 40$	$40 < a \leqslant 50$
Male %	5	20	27	13	10
Female %	8	22	29	11	8

Age, a (years)	$50 < a \leqslant 60$	$60 < a \leqslant 70$	$70 < a \leqslant 80$	$a > 80$
Male %	11	8	4	2
Female %	8	7	5	2

Display this information on a population pyramid.

4.9 Using line graphs to predict trends

You can often use data that you *do* know to estimate or predict.

■ **A continuous line graph can be used to estimate or predict from known data.**

You can read off values between the data points you know. This is called **interpolation**.

See also Chapter 7.

By continuing the line of the graph beyond the last data point you can estimate further values. This is called **extrapolation**.

Example 9

The table below shows information about the times taken for people to evacuate a building in response to a fire alarm.

Time after alarm (seconds)	Cumulative percentage evacuated
0	0
60	3
120	17
180	38
240	68
300	80
360	90

(a) Estimate how long it took 50% of the people to evacuate the building.

(b) How long after the alarm was sounded do you think the building was completely evacuated?

This information can be plotted on a line graph.

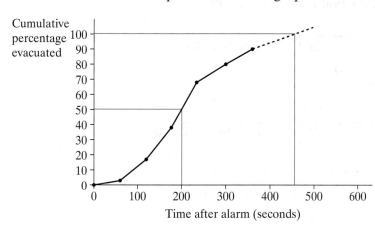

(a) You can look at the graph between values you know at 180 and 240 seconds. Read off from the graph at 50% to estimate the time as 200 seconds.

(b) You can try to carry the graph on after 360 seconds until it crosses the 100% mark at about 455 seconds.

Exercise 4H

1 The table gives information about the number of cars produced in a factory over a number of years.

Year	1990	1991	1992	1994	1995	1996	1997
Cars produced	1330	1460	1570	1740	1800	1850	1880

(a) Draw a line graph to show these data. Continue the time axis to 2000.

(b) There is a piece of information missing for 1993. Estimate how many cars were produced in the factory in 1993.

(c) Continue the graph to predict how many cars were produced in 1998, 1999 and 2000.

2 In 1990, a company aimed to have all its employees as non-smokers by the year 2005. The table shows the percentage of employees who were non-smokers at various times since 1990.

Year	1990	1992	1993	1996	1997	1999	2001
Non-smokers (%)	37	55	68	88	93	95	96

(a) Draw a line graph to show these data. Continue the time axis to 2005.

(b) Some of the years are not recorded. Estimate the percentage of non-smokers in 1991, 1994 and 1998.

(c) Predict the percentage of non-smokers for 2003.

(d) Do you think that the company will achieve its target? Explain your answer.

3 Over a period of years a group of children were regularly tested to see if they knew all their times tables. The percentage of children who knew their tables was recorded.

Age (years)	6	7	9	10	12	13	14	15
Boys (%)	1	18	35	53	65	71	78	82
Girls (%)	1	11	30	58	72	80	88	92

(a) Draw a line graph to show the data for the boys. Continue the age axis to 18 years.

(b) What percentage of 8-year-old boys would have known all their tables?

(c) Draw a line graph to show the data for girls on the same axes.

(d) Are boys or girls better at tables when they are young?

(e) Which gender is likely to be better at tables when they are adults?

(f) Predict the percentages of boys and girls who know their tables by the time they are 18.

4　The table gives information about the price of a 750 g packet of Golden Flakes every five years since 1975.

Year	1975	1980	1985	1990	1995	2000
Cost	85p	98p	£1.14		£1.56	£1.80

(a) Draw a line graph to show these data. Continue the time axis to 2005.

(b) There is a piece of information missing for 1990. Estimate what the price of a 750 g packet of Golden Flakes was in 1990.

(c) Continue the graph to predict what the price of a 750 g packet of Golden Flakes will be in 2005.

5　Andy and Sophie were both born on the same day.
Each year on their birthday they measured their heights in cm.
The table gives information about their heights at various ages.

Age (years)	6	7	8	9	10	11
Andy's height (cm)	82	98	113	129	144	156
Sophie's height (cm)	93	110	126	140	152	161

(a) Draw a line graph to show these data. Continue the age axis to 16.

(b) How tall do you think that Sophie and Andy would be at the age of 12?

(c) On which birthday do you think that Andy would be taller than Sophie?

4.10 Choropleth maps

■ **A choropleth map is used to classify regions. Regions are shaded with an increasing depth of colour.**

Example 10

The choropleth map below shows a city split into nine regions. Each region is shaded to show the percentage of people in it who are over the age of 60. The higher the percentage, the darker the region is shaded.

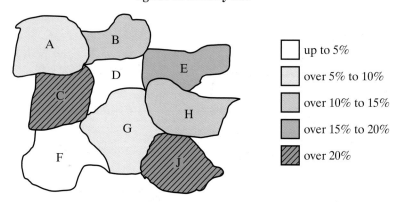

Percentage of people over the age of 60 in Buryville

up to 5%

over 5% to 10%

over 10% to 15%

over 15% to 20%

over 20%

(a) Which regions have between 10% and 15% of their population over the age of 60?

(b) Which regions have the highest proportion of people over 60 years old?

(c) Which region has a similar proportion of 60-year-olds to region D?

(a) Using the shading shown on the key, regions B and H fall in the range 10 to 15%.

(b) The highest proportion has the darkest shading. The darkest shading is used in regions C and J.

(c) Region F is shaded in the same way as region D.

Exercise 4I

1 The table below shows data for 10 regions of Heydale. It shows the percentage of houses for sale in each region which were actually sold during November.

	Region	Percentage of houses sold
A	Sanbury	12%
B	Ellerton	53%
C	Rochwood	5%
D	Barford	38%
E	Marlmore	42%
F	Radford	13%
G	Hatherford	37%
H	Birwich	22%
I	Fishton	18%
J	Carraford	32%

Use the key provided to copy and shade in the choropleth map below.

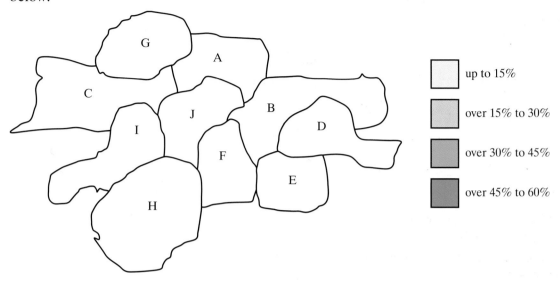

2   This choropleth map shows the percentage of adults in each region of a town who had studied in Higher Education.

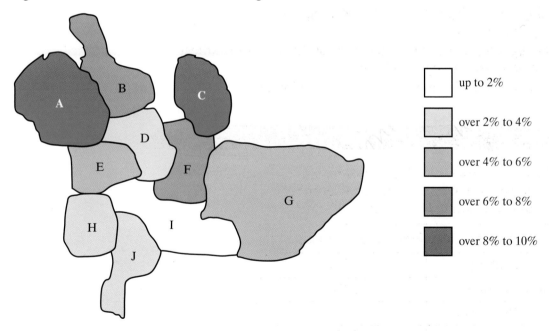

(a) Which regions have the highest proportion of adults who have studied in Higher Education?

(b) Which region has the lowest proportion of adults who have studied in Higher Education?

(c) Which region has a similar proportion of adults with experience of Higher Education to that of region F?

(d) Which regions have between 2% and 4% of adults with experience of Higher Education?

Revision exercise 4

1   Members of a youth club recorded the length of time that each member could balance a dictionary on their head. The frequency polygon shows the results for the boys.

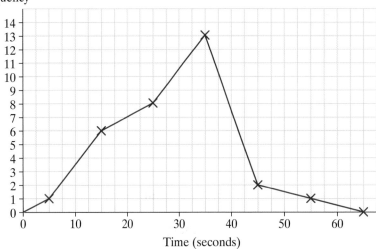

Frequency

Time (seconds)

(a) John said, 'The frequency polygon shows that 13 boys balanced the dictionary for exactly 35 seconds.'
Explain why he is wrong.

(b) Design a frequency table to show these data, using the intervals $0 < T \leqslant 10$, … etc.

(c) The table below shows similar results for the girls at the youth club.
Copy the frequency polygon above, and then draw the polygon for the girls on the same axes.

Time taken, T (seconds)	Frequency
$0 < T \leqslant 10$	3
$10 < T \leqslant 20$	10
$20 < T \leqslant 30$	14
$30 < T \leqslant 40$	5
$40 < T \leqslant 50$	2
$50 < T \leqslant 60$	1

(d) Which group were better at balancing the dictionary, the boys or the girls?
Explain your answer.

2 The table shows the percentage of the population of Fingland in each age group.

Age, a (years)	$0 < a \leqslant 10$	$10 < a \leqslant 20$	$20 < a \leqslant 30$	$30 < a \leqslant 40$	$40 < a \leqslant 50$
Male %	30	22	20	17	6
Female %	24	18	16	11	13

Age, a (years)	$50 < a \leqslant 60$	$60 < a \leqslant 70$	$70 < a \leqslant 80$
Male %	4	1	0
Female %	12	5	1

(a) Display this information on a population pyramid.

(b) Below is a list of facts about the population of Fingland. For each fact, state whether it is *true*, *false*, or whether you do not have enough information.
A. 11% of males are over 40 years old.
B. There are more males than females in the age group $0 < a \leqslant 10$.
C. In general, females have a longer life expectancy than males.
D. If I randomly selected a person in the age range $50 < a \leqslant 60$, it is more likely to be a female.
E. There is a higher proportion of males in the age group $0 < a \leqslant 10$ than that of females.
F. 36% of the population is in the age range $20 < a \leqslant 30$.

3 36 snails took part in a race. The distances in centimetres travelled by each snail were measured after half an hour. The results are shown below.

```
1.3   2.6   3.7   2.2   0.1   4.5   2     0.2   1.1
0.5   1.3   2.7   3.4   3     3     0     3.5   3.2
1.7   2.2   1.1   2.1   1.8   4.2   2.1   4.7   3.8
0.6   1.3   5.2   4.3   2.8   1.9   3.3   3.6   4.2
```

(a) Construct a frequency table using equal class intervals (e.g. $0 < d \leqslant 1$, $1 < d \leqslant 2$, ...).

(b) Construct a histogram to show these data.

(c) From the raw data, construct a stem and leaf diagram.

(d) Which of the diagrams is a better way to show the data? Explain your answer.

Summary of key points

1 When sorting continuous data into a frequency table, class intervals must be joined but not overlapping. Inequalities can be used to define the class boundaries.

2 When raw data are rounded before they are recorded, set class boundaries so that all possible values of a rounded number fit into the same class.

3 A frequency polygon is used to show the shape of a continuous frequency distribution.

4 A histogram is made up of a series of bars or rectangles. The area of each rectangle represents the frequency of a class interval.

5 For a histogram or frequency polygon, the class boundaries are written as part of a continuous scale.

6 In a histogram, frequency density = frequency ÷ class width

7 In a histogram, use frequency density for the heights of the rectangles. Make sure the rectangle spreads across the entire class interval.

8 When using frequency density the areas of the rectangles are numerically equal to the frequencies they represent.

9 Frequency = frequency density × class width

10 A stem and leaf diagram allows you to show the distribution in the same way as a frequency diagram, retaining the detail of the data.

11 A population pyramid allows you to compare aspects of a population, usually by gender.

12 A continuous line graph can be used to estimate or predict from known data.

13 A choropleth map is used to classify regions. Regions are shaded with an increasing depth of colour.

5 Summarising data: measures of central tendency and dispersion

Children are different ages, have different heights and weights, are different genders and have many other differences.

5.1 Mode, median and mean

One of the things often done in statistics is to collect data and find an average.

■ **An average is a single value used to describe a set of data.**

There are 17 young people on a school bus.

Their ages are:

12, 15, 13, 17, 10, 14, 16, 15, 16,
12, 15, 12, 13, 14, 16, 15, 11

There are many different ways of working out an average. The three most common ways are to work out the **mode**, the **median** and the **mean**.

The mode

■ **The mode is the value that occurs most often.**

Example 1

The ages of the young people on the bus are:

12, 15, 13, 17, 10, 14, 16, 15, 16,
12, 15, 12, 13, 14, 16, 15, 11

Putting these in order gives

    10, 11, 12, 12, 12, 13, 13, 14, 14,
    15, 15, 15, 15, 16, 16, 16, 17

The number in this list with the greatest frequency is **15**.
The frequency is 4.
So the mode of these numbers is 15.

The median

■ **The median is the middle number of a list of numbers after they have been put in order. (Start with the smallest and end with the biggest.)**

Example 2

The median of the ages of the 17 people on the school bus will be:

    10, 11, 12, 12, 12, 13, 13, 14, **14**, 15, 15, 15, 15, 16, 16, 16, 17

There may be an *even* number of items in the list.
Say the list was

    13, 13, 13, 14, 14, 15

In this case you can work out the median by taking the average of the two numbers in the middle.

So the median in this example is $\dfrac{13 + 14}{2} = 13\frac{1}{2}$.

The mean

A set of data can have different means. Here we are looking at the **arithmetic mean**, often called simply the mean.

■ **The mean is worked out by adding up the items then dividing that total by the number of items.**

Another kind of mean is the geometric mean (see Section 5.7).

Example 3

The ages of the young people on the school bus were

    12, 15, 13, 17, 10, 14, 16, 15, 16, 12, 15, 12, 13, 14, 16, 15, 11

The total of these ages is:

    12 + 15 + 13 + 17 + 10 + 14 + 16 + 15 + 16 + 12 + 15 + 12
    + 13 + 14 + 16 + 15 + 11 = 236

Now divide 236 by the total of 17 (this is because there are 17 numbers):

$$\frac{236}{17} = 13.88$$

Exercise 5A

1   Work out the mode, median and mean of each set of values.

   (a) 5, 7, 9, 9, 8, 7, 10, 9, 11, 12, 5, 9, 9

   (b) 4, 7, 11, 9, 12, 10, 8, 11, 14, 2, 6

   (c) 2, 7, 1, 4, 19, 11, 2, 8, 5, 6

2   The number of cars using a car park during a particular period is summarised in the stem and leaf diagram.

Number of cars

1	2 3
2	3 4 4 5
3	2 3 3 3 6 7 9
4	0 1 3 5
5	0 2

Key
2 | 3 means 23

Work out the mode, median and mean for these data.

3   The mean weight of a group of ten students is 52 kg.

   (a) What is the total weight of these students?

   One other student joins the group of ten. This student has a weight of 30 kg.

   (b) What is the mean weight of all eleven students?

4   Give an example of ten numbers which **(a)** have no mode and **(b)** are bimodal (have two modes).

5.2 Mode, median and mean of frequency data

Data are often summarised using a frequency distribution.

The mode, median and mean can be found by the same method as before but it is simpler, and quicker, to use the frequencies.

Example 4

The number of goals scored by the 21 teams in the premier division of a local soccer league last week were:

Number of goals	0	1	2	3	4	5	6
Teams	1	3	6	2	4	3	2

Find **(a)** the mode, **(b)** the median, **(c)** the mean.

The table shows that 1 team scored 0 goals, 3 teams scored 1 goal, 6 teams scored 2 goals, etc.

(a) The mode is the number of goals scored by most teams.
So the mode is 2 because there were six teams that scored
2 goals.

(b) The median number of goals is the middle number.
To find this it is often sensible to write the list in terms of a
cumulative frequency or, as it is sometimes known, a **running
total**.

To find a running total up to,
say, 4 goals, add the number
of teams scoring
4 goals to the number of
teams scoring less than
4 goals.
So, up to
4 = 1 + 3 + 6 + 2 + 4 = 16
= up to 3 + no. scoring 4

Number of goals	0	up to 1	up to 2	up to 3	up to 4	up to 5	up to 6
Teams	1	4	10	12	16	19	21

You want the median number. You can work out which term you
want by adding 1 to the total number of teams and dividing by 2:

$$\frac{21 + 1}{2} = 11$$

The term you want is the 11th out of 21. The median is 3, as this
is the number of goals scored by the 11th team in the list.

In some cases there will be an even number of teams (or
whatever).
Imagine that there had been 20 teams in the league and that the
distribution of goals was like this:

Number of goals	0	1	2	3	4	5	6
Teams	1	3	6	3	4	2	1

There is no number exactly in the middle, but the middle
numbers are 10 and 11.
The distribution looks like this:

$$\frac{20 + 1}{2} = 10.5$$

Middle numbers are the 10th
and 11th.

0 1 1 1 2 2 2 2 2 2 3 3 3 4 4 4 4 5 5 6

The two numbers in the middle are 2 and 3.
The median is then simply the mean of 2 + 3:

$$\text{median} = \frac{2 + 3}{2} = 2\tfrac{1}{2}$$

(c) To find the mean number of goals in the original example, list
them as:

0 1 1 1 2 2 2 2 2 2 3 3 4 4 4 4 5 5 5 6 6

then add these numbers together as

$$\Sigma(0 + 1 + 1 + \dots + 6) = 64$$

Σ means 'sum of'

and divide this total by 21 (the number of teams).

$$\frac{64}{21} = 3.048 \text{ (correct to three decimal places)}$$

So the mean = 3.048 goals.

Normally you can do the same thing by extending the table:

Number of goals, x	Number of teams, f	$f \times x$
0	1	$1 \times 0 = 0$
1	3	$3 \times 1 = 3$
2	6	$6 \times 2 = 12$
3	2	$2 \times 3 = 6$
4	4	$4 \times 4 = 16$
5	3	$3 \times 5 = 15$
6	2	$2 \times 6 = 12$
		Total = 64

$$\text{Mean} = \frac{64}{21} = 3.048 \text{ (correct to three decimal places)}$$

Exercise 5B

1   The following frequency distribution summarises the number of cartons of yoghurt sold by a shopkeeper during January.

Number of cartons	10	11	12	13	14	15	16	17
Number of days	1	3	5	10	6	3	2	1

Find for these numbers of cartons of yoghurt:
(a) the mode, (b) the median, (c) the mean.

2   The following frequency distribution provides information about the number of times each student in a Year 11 class is late during a term.

Times late	2	3	4	5	6	7	8
Number of students	3	12	6	4	7	2	1

For these data, work out:
(a) the mode, (b) the median, (c) the mean.

3   During two consecutive months a gardener wrote down the temperature, in °C, at the same time each day. The results he obtained are provided here.

Temperature (°C)	18	19	20	21	22	23
Number of days	5	8	19	14	12	3

Find for these temperatures:
(a) the mode, (b) the median, (c) the mean.

4 The frequency distribution below shows the ratings of a disco by a random sample of 30 students.
A capital A means they enjoyed it very much.
A capital E means that they did not enjoy it at all.

Rating	A	B	C	D	E
Number of students	6	13	10	7	4

(a) Work out (i) the mode, (ii) the median for this distribution.

(b) Explain why you cannot write down the mean of this distribution.

5.3 Mode, median and mean of grouped data

In many cases data are summarised in a grouped frequency table. The table below shows information about the time, to the nearest hour, that a group of youngsters spent watching television one week.

See page 31.

Mode or modal class of grouped data

Number of hours	Number of youngsters
$0 \leqslant$ hours < 5	3
$5 \leqslant$ hours < 10	7
$10 \leqslant$ hours < 15	10
$15 \leqslant$ hours < 20	4

- The advantage of using such a table is that data can be summarised quickly.
- The disadvantage of using such a table is that only estimates of details such as the mode, median and mean can be given.

We cannot give the mode but we can say that the **modal class interval** is $10 \leqslant$ hours < 15 because this is the class interval with the most observations in it. Of course the modal class interval will change if we change the class intervals.

■ **When information is presented in a table of grouped data, the modal class interval is the class interval with the largest frequency.**

Median of grouped data

When you are dealing with grouped data you will never be able to state what the median is, but you can usually estimate it.

For the data connecting the number of hours spent watching the television with the number of youngsters, we have:

No. of hours	0 to <5	5 to <10	10 to <15	15 to <20
No. of youngsters	3	7	10	4

There were 24 youngsters in the sample, so the middle one is the average of the 12th and 13th, that is, $12\frac{1}{2}$. We actually want the number of hours of TV watched by the $12\frac{1}{2}$th person.

Exactly 10 of the youngsters watch TV for less than 10 hours, so we want $2\frac{1}{2}$ out of the next class interval, which is 10 to less than 15.

$15 - 10 = 5$

$2\frac{1}{2} = \frac{1}{4}$ of 10 (there are 10 youngsters in this class interval)

$\frac{1}{4}$ of $5 = 1\frac{1}{4}$

So the estimated median $= 10 + 1\frac{1}{4} = 11\frac{1}{4}$ hours.

You could also estimate the median by using a cumulative frequency diagram.
The cumulative frequencies are:

Time	Cumulative frequency
Less than 5 hours	3
Less than 10 hours	10
Less than 15 hours	20
Less than 20 hours	24

$(3 + 7 = 10)$
$(10 + 10 = 20)$
$(20 + 4 = 24)$

Use the upper bounds of the classes to plot the points.

Plot these points and find the mid-point or the $12\frac{1}{2}$th point.

The points could also have been joined up by straight lines to create a **cumulative frequency polygon**.

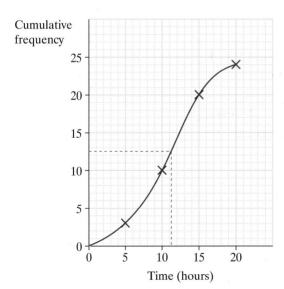

Median = just over 11
$\approx 11\frac{1}{4}$

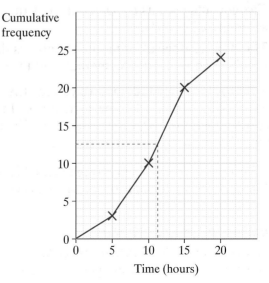

Median = just over 11
$\approx 11\frac{1}{4}$

Whichever method you use you will get an *estimate* for the median.

■ **The median can be found using the mid-point of a cumulative frequency diagram or polygon.**

We now have three values of the median. None of them is accurate; an estimate is as good as we can obtain:

Mean of grouped data

The best we can do is to work out an estimate for the mean value of any grouped data. We shall use the same example and work out an estimate for the mean number of hours of TV watched by 24 youngsters. The results, when grouped, were:

Number of hours	Number of youngsters
$0 \leqslant \text{hours} < 5$	3
$5 \leqslant \text{hours} < 10$	7
$10 \leqslant \text{hours} < 15$	10
$15 \leqslant \text{hours} < 20$	4

Start by replacing each class interval with its mid-point. If you know that 3 youngsters watch TV for between 0 and 5 hours then you can estimate that each watches it for an average of $2\frac{1}{2}$ hours.

Mean number of hours	Number of youngsters
$2\frac{1}{2}$	3
$7\frac{1}{2}$	7
$12\frac{1}{2}$	10
$17\frac{1}{2}$	4

To find the mid-point of the class interval, add the upper and lower bounds and divide by 2.

e.g. $\dfrac{0+5}{2} = 2\frac{1}{2}$

Then the mean becomes:

$$\frac{3 \times 2\frac{1}{2} + 7 \times 7\frac{1}{2} + 10 \times 12\frac{1}{2} + 4 \times 17\frac{1}{2}}{24}$$

which is

$$\frac{7\frac{1}{2} + 52\frac{1}{2} + 125 + 70}{24} = \frac{255}{24} = 10.625 \text{ hours}$$

So an estimate for the mean of these grouped data is 10.625 hours.

■ **An estimated mean can be found from a grouped set of data, using the formula**

$$\text{Mean} = \frac{\Sigma(f \times \text{mid-point})}{\Sigma f}$$

where Σ is 'sum of' and f is the frequency.

Exercise 5C

1 The speeds of some cars on a motorway are given in the frequency table.

Speed (mph)	Number of cars
20 < speed ≤ 30	3
30 < speed ≤ 40	10
40 < speed ≤ 50	17
50 < speed ≤ 60	30
60 < speed ≤ 70	35
70 < speed ≤ 80	5

Use this information to work out:

(a) the modal group for the speed of the cars used in this survey,

(b) an estimate for the median speed of these cars,

(c) an estimate for the mean speed of these cars.

2 The ages of some people watching a film are given in this frequency table.

Age	Number of people
10 ≤ age < 20	4
20 ≤ age < 30	15
30 ≤ age < 40	11
40 ≤ age < 50	10

Work out:

(a) the modal age of the people watching the film,

(b) an estimate for the median age of the people watching the film,

(c) an estimate for the mean age of the people watching the film.

5.4 Transformations

Sometimes it is useful to make the numbers being used easier by **scaling**.

Example 5

The ages, in years, of some people in an a retirement home are:

    76   81   73   92   83

Work out the mean age.

The median of 76, 81, 73, 92, 83 is 81

You could add these five numbers and then divide this total by 5. But you can make the arithmetic easier by subtracting 70 from each number:

6   11   3   22   13

The mean of these five numbers is

$$\frac{6 + 11 + 3 + 22 + 13}{5} = 11$$

The median of 6, 11, 3, 22, 13 is 11

Now all you need to do is add the 11 to 70 (the number you reduced them all by) to give the mean as

mean = 70 + 11 = 81

The median = 70 + 11
= 81

Example 6

Find the mean of these numbers:

1.04, 1.09, 1.03, 1.12, 1.10, 1.04

Take away 1, to give

0.04   0.09   0.03   0.12   0.10   0.04

then multiply by 100 to give

4   9   3   12   10   4

The mean of these numbers is

$$\frac{4 + 9 + 3 + 12 + 10 + 4}{6} = \frac{42}{6} = 7$$

The mode here is 4.
So the mode of the original data is 1.04.

The mean can now be found by reversing what you have done:
divide 7 by 100 to give 0.07,
then add on 1 to give the mean as 1.07.

Exercise 5D

1   Find the mean of the seven numbers by scaling (−3000).

3003   3005   3001   3010   3004   3009   3002

2   Work out the mean of the set of numbers by scaling (−2 then ×100).

2.14   2.11   2.20   2.18   2.12   2.13   2.17   2.18

5.5 Deciding which average to use

Whenever you use the word 'average' you should be clear what the word means. The mode, median and mean are all 'averages' but each one has a very different meaning.

Use this table to help you decide when to use the mode or the median or the mean.

Average	Advantages	Disadvantages
Mode	Easy to find. Can be used with any type of data. Not affected by open-ended or extreme values. The mode will be a data value.	Mathematical properties are not very useful. There is not always a mode.
Median	Easy to calculate. Unaffected by extremes.	Mathematical properties are not very useful. Not always a given data value.
Mean	Uses all the data. Mathematical properties are well known and useful.	Always affected by extreme values. Can be distorted by open-ended classes.

Example 7

The salaries of seven people who work for a small company are

£12 000	£18 000	£120 000	£28 000
£32 000	£22 000	£30 000	

(a) Work out (i) the mean salary, (ii) the median salary.

(b) Which of these two averages is the most typical of a person's earnings?

(c) Why is it not possible to work out the mode of the salaries?

(a)   (i) The mean is

$$\frac{12 + 18 + 120 + 28 + 32 + 22 + 30}{7} = 37.428571$$

i.e. £37 429

  (ii) The median is 12, 18, 22, **28**, 30, 32, 120,

i.e. = £28 000

(b) The median is more typical because the single salary of £120 000 affects the mean and only 1 out of 7 earns this level of money.

(c) We cannot give a mode because all the values are different.

Exercise 5E

1   (a) Find the median of the numbers

8, 10, 10, 14, 16, 17, 100

  (b) Which of the 'averages' would you chose to describe these numbers?
Give a reason for your answer.

2 (a) Work out the mean of the ten numbers

18, 22, 17, 26, 18, 27, 25, 18, 17, 20

(b) Why is the mode not a good choice as a measure of the average of this set of values?

3 Explain in a few words how you could work out the average price of a car that is sold by a garage.

> Mention the words mode, median and mean in your explanation.

4 Last April, the garage sold five different types of cars. The numbers sold were:

Car	Prestige	Sports	Ordinary	Coupe	4 by 4
Number sold	12	8	23	2	5

(a) Find the modal type of car sold.

(b) Why is the mode appropriate for these data?

5 Twenty students were asked to name their favourite colour. Here are their results

red	blue	blue	green	red
red	yellow	red	blue	pink
blue	black	red	red	blue
purple	blue	red	red	red

(a) State clearly, with reasons, which is the best average to use for the favourite colour of the students.

(b) Why can you not state the mean of the favourite colours?

6 Jamie and Suzanne are taking a walking holiday in Spain. They record the temperature at mid-day for a week of their holiday.

Day	Mon	Tues	Wed	Thur	Fri	Sat	Sun
Temperature (°C)	24	23	25	22	27	30	28

(a) Work out the mean mid-day temperature during the week.

(b) Say why you could not find the mode of these mid-day temperatures.

(c) Would the mean be a reasonable average mid-day temperature?
Give a reason for your answer.

5.6 The weighted mean

In GCSE coursework, a candidate's final, overall percentage is worked out by using the rule:

$$\frac{40 \times \text{Paper 1 mark} + 40 \times \text{Paper 2 mark} + 10 \times \text{Paper 3 mark} + 10 \times \text{Paper 4 mark}}{40 + 40 + 10 + 10}$$

A candidate scored the following marks:

    Paper 1: 62%   Paper 2: 38%   Paper 3: 58%   Paper 4: 39%

So for this candidate the overall or final mark would be

$$\frac{40 \times 62 + 40 \times 38 + 10 \times 58 + 10 \times 39}{40 + 40 + 10 + 10}$$

40, 40, 10 and 10 being the weightings given to the four papers.

$$\frac{2480 + 1520 + 580 + 390}{100} = \frac{4970}{100} = 49.7\%$$

This is an example of working out a **weighted mean**.

Example 8

A sample of 3 children had a mean height of 0.97 metres, a second sample of 7 children had a mean height of 1.06 metres and a final sample of 5 children had a mean height of 1.12 metres.
Work out the mean height of all 15 children.

$$\text{mean height} = \frac{3 \times 0.97 + 7 \times 1.06 + 5 \times 1.12}{3 + 7 + 5}$$

$$= \frac{15.93}{15}$$

$$= 1.062 \text{ metres}$$

Exercise 5F

1   In an examination, Paper 1 is worth 0.35 of the total, Paper 2 is worth 0.4 of the total and Coursework is worth 0.25 of the total.
The marks for Jimmy and Sumreen were

	Paper 1	Paper 2	Coursework
Jimmy	46	62	55
Sumreen	56	56	65

    (a) Work out each student's final mark.

    (b) Comment on the final marks.

2   The weightings used to work out a final average mark in an examination were 0.3, 0.4, 0.2 and 0.1. Owen scored marks of 62, x, 44 and 58.
The overall pass mark was 55.
Work out the value of x, given that Owen just passed the examination.

5.7 The geometric mean

On page 110 you saw that to work out the *arithmetic* mean you add the numbers together and then divide by the number of them.

■ **To work out the geometric mean of *n* numbers, multiply the numbers together and then take the *n*th root of the product.**

Geometric mean = $\sqrt[n]{x_1 \times x_2 \times \ldots \times x_n}$

Example 9

Work out the geometric mean of the numbers

   3, 7, 4, 8

The geometric mean is $\sqrt[4]{(3 \times 7 \times 4 \times 8)}$

$$= \sqrt[4]{672}$$

$$= 5.09 \text{ (to two decimal places)}$$

(Note that the arithmetic mean is $22/4 = 5.5$)

The geometric mean is usually used in situations where relative changes are involved.

Example 10

Mrs Jenkins invests £4000. The amount that it is worth over the next six years is:

Year 1	Year 2	Year 3	Year 4	Year 5	Year 6
£5000	£7000	£8000	£9000	£12 000	£15 000

Work out the geometric mean of the year by year increases.

The relative year by year increases are:

$$\frac{5000}{4000} \qquad \frac{7000}{5000} \qquad \frac{8000}{7000} \qquad \frac{9000}{8000} \qquad \frac{12\,000}{9\,000} \qquad \frac{15\,000}{12\,000}$$

and the geometric mean of these six numbers is

$$\sqrt[6]{(1.25 \times 1.40 \times 1.14 \times 1.125 \times 1.33 \times 1.25)} = 1.245$$

Exercise 5G

1   Work out the geometric mean of
    (a) 3, 6, 2, 5, 8
    (b) 1.24, 1.31, 1.26, 1.20, 1.27, 1.22, 1.32

2   (a) Work out the geometric mean of 13, 12, 0, 14, 15, 16
    (b) Why will it be a silly measurement of an average value?

3 The turnover of a particular company changed from an initial value of £48 000 to £55 000, £67 000 and £90 000. Work out the geometric mean of the year by year increases.

5.8 Index numbers

An index number is an indicator of relative change.

■ **An index number shows the rate of change in quantity, value or price of an item over a period of time.**

An index number is a percentage which gives you the size or the value of a quantity relative to a standard number, or base.

$$\text{Index number} = \frac{\text{quantity}}{\text{quantity in base year}} \times 100$$

Simple index numbers

An example of an index number, perhaps the simplest of all, is a **price relative**.

■ **A price relative shows the change in price over time. It is calculated as a percentage of its value at a given base year.**

$$\textbf{Price relative} = \frac{\textbf{price}}{\textbf{price in base year}} \times \textbf{100}$$

Example 11

The value of a house is given for the years from 2000 to 2003.

Value (£)	£80 000	£88 000	£112 000	£110 000
Year	2000	2001	2002	2003

Work out the price relative using 2000 as the base year.

Price relative for 2001 is $\dfrac{\text{price in 2001}}{\text{price in 2000}} \times 100 = \dfrac{88\,000}{80\,000} \times 100 = 110$

Price relative for 2002 is $\dfrac{112\,000}{80\,000} \times 100 = 140$

Price relative for 2003 is $\dfrac{110\,000}{80\,000} \times 100 = 137.5$

The value of the index number (price relative) for each chosen base year will always be 100.

Example 12

A product's price index numbers for the three years from 2001 to 2003 are

Year	2001	2002	2003
Index	100	106	124

(a) Work out the price of the product in 2003 given that its price in 2002 was £300.

The new base year is set as 2002.

(b) Find the new index number for 2003.

(a) The 2003 price = price in 2001 × 124
The 2002 price = price in 2001 × 106
So 300 = price in 2001 × 106
Hence

$$\text{price in 2001} = \frac{300}{106}$$

the 2003 price = price in 2001 × 124

$$= \frac{300}{106} \times 124$$
$$= £350.94$$

(b) Index number for 2003

$$= \frac{\text{old index for 2003}}{\text{old index for 2002}} \times 100$$
$$= \frac{124}{106} \times 100$$
$$= 117$$

Chain base index numbers

To work out how the price of an item has changed over one year, use the previous year as the base year. This is called a **chain base index number**.

■ **A chain base index number tells you the annual percentage change. It is found by using the previous year as the base year and then working out the relative value of an item.**

Example 13

The value of a car is given in this table.

Age (years)	0	1	2	3
Value (£)	10 000	8300	7000	5800

Find the chain base index numbers for each year.

The chain base index numbers are:

Year 1 $\dfrac{8300}{10\,000} = 0.83$

Year 2 $\dfrac{7000}{8300} = 0.843$

Year 3 $\dfrac{5800}{7000} = 0.829$

Weighted index numbers

Brass is a mixture of 70% copper and 30% zinc. Clearly, if the price of producing either copper or zinc changes then so does the price of producing brass. The price of brass will be affected more by the price of copper than by the price of zinc.

The price index for brass will reflect the different proportions of the two elements, copper and zinc, used in its manufacture. We need to set up a **weighted index number** to help to calculate any changes in the cost.

■ **To calculate a weighted index number you need to:**
 – calculate the index number for every element, and then
 – find the weighted average of all of the elements.

Example 14

Brass is made using a mixture of copper and zinc in the ratio of 70% to 30%.
The table gives the price of copper and zinc, per tonne in the two years 1996 and 2003.

Year	1996	2003
Copper (£)	1230	1580
Zinc (£)	780	790

Using 1996 as the base year, find the weighted average for the change in cost of brass. Comment on the result.

The index for each element will be $\dfrac{\text{price in year 2003}}{\text{price in year 1996}} \times 100$

So, for copper the index is $\dfrac{1580}{1230} \times 100 = 128.46$

and for zinc it is $\dfrac{790}{780} \times 100 = 101.28$

So the index for copper is $128.46 \times 70 = 8992.2$
And the index for zinc is $101.28 \times 30 = 3038.4$

Thus the weighted index for brass is

$\dfrac{8992.2 + 3038.4}{70 + 30} = \dfrac{12\,030.6}{100} = 120.306$ or 120 approximately.

This weighted index tells you that the average price of the raw materials rose by 20.306% or just over 20%, and this was mainly due to the rise in the price of copper.

Retail price index

The **retail price index**, or RPI, is a type of weighted index which is used to show or monitor changes and make comparisons. The aim of the RPI is to show changes in the cost of living of an average family or person.

■ **The retail price index is a weighted mean of the price relatives of goods and services. The weightings are chosen in such a way as to show the spending habits of an average household.**

For just over forty years the retail price index has been calculated by revising the weightings each year, based on the previous three years.

The items used to calculate the RPI change over time. For example, nowadays, most families have a home computer which is a very different expense from a few years ago.

Example 15

The Jordan family home was valued in July 2002 as being worth £240 000.
Mr and Mrs Jordan bought the house in 1982 for £40 000. They have made no essential improvements to their house during the 20 years they have owned it.
The value of houses has stayed in line with the increase in basic costs.
If the base year for the RPI was 1982, work out:

(a) the RPI in 2002,

(b) the yearly inflation rate if this is assumed to be constant.

(a) The RPI $= \dfrac{\text{price of house in 2002}}{\text{price of house in 1982}} \times 100$

$$= \frac{240\,000}{40\,000} \times 100 = 600$$

(b) If the yearly inflation rate is $x\%$, then:

$$\left(1 + \frac{x}{100}\right)^{20} \times 100 = 600$$

$$\left(1 \times \frac{x}{100}\right)^{20} = 6$$

$$1 + \frac{x}{100} = \sqrt[20]{6} = 1.093\,723\,548$$

$$\frac{x}{100} = 0.093\,723\,548$$

$$x = 9.37$$

This means that the annual rate of inflation, x, would be just over 9%. The government tries to keep inflation much lower than 9% each year. The normal rate of inflation is around 2 to $2\frac{1}{2}\%$.

Exercise 5H

1   In year 2000 the price of a litre of unleaded petrol was 84p. In 2003 the price of a litre of unleaded petrol was only 77p. Using the year 2000 as the base year, calculate the index number (price relative) for a gallon of unleaded petrol.

2   The value of Joy's flat is shown for the years between 1998 and 2003.

Value (£)	75 000	82 000	90 000	102 000	110 000	134 000
Year	1998	1999	2000	2001	2002	2003

(a) Calculate the chain base index numbers for the value of Joy's flat over these six years.

Joy bought a new car at the same time that she bought the house. Here are the values of the car during the same period of time.

Value (£)	12 000	9600	7700	6200	5000	4100
Year	1998	1999	2000	2001	2002	2003

(b) Calculate the chain base index numbers for the value of Joy's car over these same six years.

(c) Comment on your findings.

3   Working as a group of at least four people, conduct a survey of expenditure.
Decide amongst yourselves how you will classify spending habits. Find the average prices of the goods you decide to survey over the last four years.
Calculate the retail price index for your group.
Write a report of your findings which you can present to those who pay money to you, such as your parents or the owner of the shop where you work on Saturday.

5.9 Range, quartiles and percentiles

As well as an average, such as the mode, median or mean, you need a measure of the spread of the data about the average if you wish to describe it more fully.

The range

■   **Range = largest value − smallest value**

The range is a very crude measure of spread because it compares only the largest and the smallest values of the data.

Example 16

During a survey of the late arrivals at a school, the headteacher recorded the number of students who arrived late for the start of the day. The minimum number of late arrivals was 8 whilst the maximum number of late arrivals was 25.
Work out the range of these recorded late arrivals.

Range = maximum number − minimum number
$$= 25 - 8$$
$$= 17 \text{ students}$$

Example 17

The speeds of cars on a motorway were recorded by the police.
The table of their final results was:

Speed, s (mph)	Number of cars
$20 < s \leqslant 30$	2
$30 < s \leqslant 40$	14
$40 < s \leqslant 50$	29
$50 < s \leqslant 60$	22
$60 < s \leqslant 70$	13

Estimate the range of the speeds.

Range = largest value − smallest value

The largest value appears to be 70, but speed is continuous so the largest speed is 70.5 mph.
Similarly the smallest speed is 20.5 mph.

$20 < s$ so the smallest value of s is 20.5. Speeds between 20 and 20.5 would have been rounded to 20.

So the range = $70.5 - 20.5 = 50$ mph

Quartiles and inter-quartile range

The **lower quartile** is a value such that one quarter of the values are less than or equal to it.
The **upper quartile** is a value such that three quarters of the values are less than or equal to it.
The quartiles split data into four equal parts.

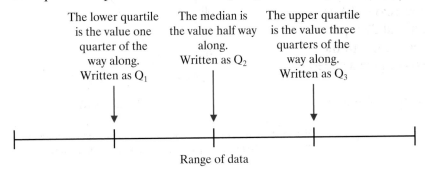

The lower quartile is the value one quarter of the way along. Written as Q_1

The median is the value half way along. Written as Q_2

The upper quartile is the value three quarters of the way along. Written as Q_3

Range of data

■ **When n data values are written in ascending order:**
  - **the lower quartile, Q_1, is the $\frac{1}{4}(n + 1)$th value**
  - **the median, Q_2, is the $\frac{1}{2}(n + 1)$th value**
  - **The upper quartile, Q_3, is the $\frac{3}{4}(n + 1)$th value.**

Q_1 is often referred to as LQ, Q_2 as the median and Q_3 as UQ

If you get a non-integer for $\frac{1}{4}(n + 1)$ or $\frac{3}{4}(n + 1)$, round up;
e.g. if $\frac{1}{4}(n + 1) = 2\frac{3}{4}$ find the third value.
If you get a non-integer for $\frac{1}{2}(n + 1)$, take the two integers on either
side; e.g. if $\frac{1}{2}(n + 1) = 4\frac{1}{2}$ use the fourth and fifth values.

A crude measure of the spread is the **inter-quartile range**, or **IQR**,
where

■ **Inter-quartile range = upper quartile − lower quartile.**

Example 18

(a) Find the upper and lower quartiles of the set of data
$$7, 9, 13, 5, 6, 12, 3$$
(b) Find the inter-quartile range for the data.

(a) The first thing you need to do is order the data:
$$3, 5, 6, 7, 9, 12, 13 \text{ and note that } n = 7.$$
Then the lower quartile or Q_1 is $\frac{1}{4}(7 + 1) = 2$ or the 2nd value,
which is 5.
So $Q_1 = 5$.
$Q_3 = \frac{3}{4}(7 + 1) = \frac{3}{4}(7 + 1) = 6$ or the 6th value, which is 12.
(b) The inter-quartile range is $IQR = Q_3 - Q_1 = 12 - 5 = 7$

Example 19

Here is a cumulative frequency graph showing
the speed of 100 cars on a road.

Use the graph to find approximate values of

(a) the lower quartile speed,
(b) the upper quartile speed,
(c) the inter-quartile speed.

(a) To find the lower quartile speed, go up to 25
on the vertical axis, (25 is $\frac{1}{4}$ of 100), across to
the curve and down to the horizontal axis.
Then read off the x value: about 32 mph.
(b) Go to 75 on the vertical axis, across to the
curve and down to the x-axis to give a speed
of 56 mph.
(c) The inter-quartile range is $56 - 32 = 24$ mph.

When the data are continuous and can take values between
whole numbers, use $\frac{1}{4}n$, $\frac{1}{2}n$ and $\frac{3}{4}n$ to find the quartiles.

Percentiles and deciles

■ A set of data is subdivided into 100 equal parts to form percentiles.

■ When the set of data is subdivided into 10 equal parts, these are called deciles.

Example 20

From the cumulative frequency graph in Example 19, find
(a) the 35th percentile,   (b) the 80th percentile,
(c) the 80th–35th percentile range,   (d) the 6th decile.

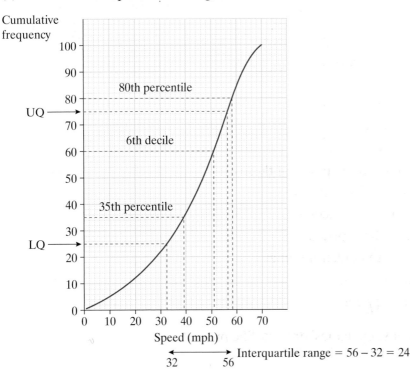

The 6th decile is the same as the 60th percentile

(a) The 35th percentile is approximately 39 mph.
(b) The 80th percentile is approximately 58 mph.
(c) The 80th–35th percentile range is
$$58 - 39 = 19 \text{ mph}$$
(d) The 6th decile is approximately 51 mph.

Exercise 5I

1   Find the range, lower quartile, upper quartile and inter-quartile range of the values
    (a) 6, 3, 8, 2, 9, 5, 10
    (b) 21, 16, 72, 40, 67, 65, 55, 34, 17, 48, 32, 19, 44, 61, 73
    (c) 8, 2, 9, 6, 7, 10, 12, 13, 5, 12, 10, 8, 10, 4

2 The ages, in years, of a group of 60 people are shown in the
  table below.

Age (years)	$10 < a \leqslant 20$	$20 < a \leqslant 30$	$30 < a \leqslant 40$	$40 < a \leqslant 50$	$50 < a \leqslant 60$
Number	4	12	22	19	3

  (a) Draw a cumulative frequency curve.

  (b) Find the lower quartile, the upper quartile and the inter-
    quartile range.

3 The prices of 200 second-hand cars are shown in the frequency
  table.

Price, x (£1000s)	$1 < x \leqslant 2$	$2 < x \leqslant 3$	$3 < x \leqslant 4$	$4 < x \leqslant 5$	$5 < x \leqslant 6$
Frequency	10	32	95	51	12

  (a) Draw a cumulative frequency curve.

  (b) Find:

    (i) the median price of a car,

    (ii) the upper quartile price of a car,

    (iii) the lower quartile price of a car,

    (iv) the inter-quartile range in the prices of these cars.

  (c) Work out the approximate values of:

    (i) the 85th percentile of these prices,

    (ii) the 35th percentile of these prices,

    (iii) the 85th percentile–35th percentile range,

    (iv) the 4th decile of these prices,

    (v) the 8th decile of these prices.

4 Angela is studying the cost of some ladies' dresses. She presents
  the results of her survey as a stem and leaf diagram.

```
1 | 3  5  8
2 | 1  2  2  2  3  5  6  6  6  7  8  9  9  9
3 | 2  2  2  2  2  3  7  7  9  9
4 | 0  2  3  4  4  8  7  8  8
5 | 1  2  2  6  7  8  8  9
6 | 2  3  8  8  8
7 | 5  5  9
```

Key
$2 \mid 1 = £21$

  (a) Write down the median of these costs.

  (b) Draw a cumulative frequency diagram to show these prices.

  (c) Use your cumulative frequency diagram to work out
    estimates of

Use the classes
$10 \leqslant c < 20$
$20 \leqslant c < 30$
etc.

    (i) the lower and upper quartiles,

    (ii) the 6th decile of the prices,

    (iii) the 15th percentile of the prices.

Box and whisker diagrams

Once you know the quartiles, the median and the range you have sufficient information to be able to highlight some of the main features of a distribution.

■ **A box and whisker diagram can be drawn to represent important features of the data, e.g. to show the maximum and minimum values, the median and the upper and lower quartiles.**

A **box and whisker diagram** or box plot looks like this:

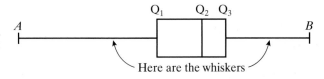

The two values A and B are the minimum and maximum values of the data respectively.

- If the median is closer to the upper quartile than it is to the lower quartile then the data are said to be **negatively skewed**.

- If the median is closer to the lower quartile than it is to the upper quartile then the data are said to be **positively skewed**.

- If the median is equidistant from both the upper and lower quartiles then it is likely that the distribution is **symmetrical**.

The diagram above shows negative skew.

> For comparitive purposes, two box and whisker diagrams may be drawn, one above the other. Use the same scale for both, e.g.
>
> Scale

Example 21

The heights, in centimetres, of 15 students are given below:

163, 170, 182, 164, 155, 172, 177, 184, 190, 148, 193, 185, 176, 158, 166

(a) Find the median height, the quartiles and the inter-quartile range.
(b) Draw a box and whisker diagram.
(c) Describe the skewness of this distribution.

(a) First put the list of 15 numbers in order:
148, 155, 158, 163, 164, 166, 170, **172**, 176, 177, 182, 184, 185, 190, 193
The number in bold, i.e. 172, is the median.
The lower quartile, or Q_1, will be the $\frac{1}{4}(15 + 1)$th value
= 4th value = 163
The upper quartile, or Q_2, will be the $\frac{3}{4}(15 + 1)$th value
= 12th value = 184

(b) The box and whisker diagram will now be:

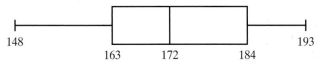

(c) You can see that the median, or Q_2, is closer to Q_1 than it is to Q_3. This means that we can say that the distribution is positively skewed.

Outliers

If in the survey described in Example 21, one girl had written that her height, in centimetres, was 52 cm, this would be incorrect! It would be an **outlier**, an exceptional case.

■ **An outlier is any value which is 1.5 (or more) times the inter-quartile range below the lower quartile or 1.5 (or more) times the inter-quartile range above the upper quartile.**

Any outliers should be marked on the box and whisker diagram, but instead of the whiskers showing the whole range they should show only:

● down to the lowest value which is not an outlier

● up to the highest value that is not an outlier.

Example 22

Represent these numbers, written in order, using a box and whisker diagram.

22, 30, 34, 35, 35, 35, 36, 37, 37, 38, 39, 39, 40, 40, 41, 42, 42, 48, 50

Find the median and the quartiles first:

Median = 38,   Lower Quartile = 35,
Upper Quartile = 41,   IQR = 41 − 35 = 6

1.5 × IQR = 1.5 × 6 = 9

Every value within LQR − 9 to UQR + 9 is not an outlier.

So the non-outliers are values within the range

(35 − 9) to (41 + 9)

i.e. within the range 26 to 50.

So the outlier is 22.

The lowest value that is not an outlier is 30.

The highest value that is not an outlier is 50.

So the box and whisker diagram looks like this:

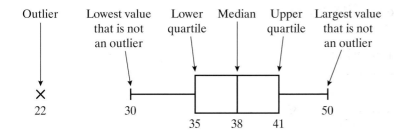

Exercise 5J

1 Class 11A at Mayfield High School took a test in Statistics. The marks obtained are given in the table below.

Girls	62	47	72	50	68	81	45	52	77	46	45	57	80	41	63
Boys	52	89	46	19	22	34	56	97	44	38	47	99	23	20	65

(a) Work out the inter-quartile range for
   (i) the girls' marks,
   (ii) the boys' marks.

(b) Work out the range of
   (i) the girls' marks,
   (ii) the boys' marks.

(c) Draw the box and whisker diagrams for both the girls' marks and for the boys' marks on the same axes.

(d) Make comparisons of the skewness of the marks for the girls and for the boys.

2 Here are the ages of a group of 27 people:

23, 31, 14, 27, 32, 34, 28, 29, 41, 28, 37, 27, 28, 41, 75,
26, 31, 42, 34, 25, 26, 30, 40, 27, 52, 36, 20

(a) Identify any outliers.

(b) Draw a box and whisker diagram for these data.

3 Here is a set of marks scored by the students in Class 10K for a test on spelling.

58, 12, 34, 42, 28, 33, 67, 63, 51, 47, 24, 32, 43, 30, 21, 25
32, 44, 34, 32, 76, 45, 55, 43, 52, 38, 44, 36, 28, 32, 45

(a) Find the median, lower quartile and upper quartile for these marks.

(b) Identify any outliers.

(c) Draw a box and whisker diagram.

4 Here is a box and whisker diagram:

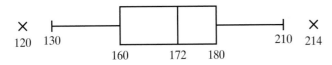

(a) Give the lower and upper quartiles, and the median.

(b) State what the outliers are.

5.10 Using a cumulative frequency polygon

Measures of spread can be readily found if you first draw a cumulative frequency polygon.

Example 23

The table below shows the heights of 50 students in Year 10 at a school.

Height, x (cm)	Number of students
$140 < x \leqslant 150$	6
$150 < x \leqslant 160$	9
$160 < x \leqslant 170$	15
$170 < x \leqslant 180$	10
$180 < x \leqslant 190$	7
$190 < x \leqslant 200$	3

(a) Draw a cumulative frequency polygon for these data.

(b) Estimate, from the graph:
    (i) the median,
    (ii) the lower quartile,
    (iii) the upper quartile,
    (iv) the inter-quartile range.

(c) Draw a box and whisker diagram.

(d) Estimate from the graph:
    (i) the 3rd percentile,
    (ii) the 4th decile.

(a)

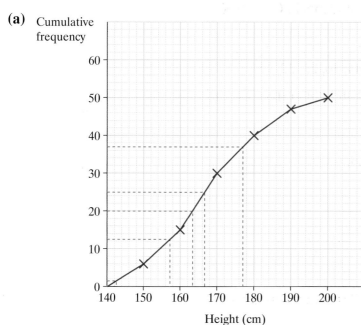

(b) The estimates are

    (i) the median = 167 cm,

    (ii) the lower quartile = 157 cm,

    (iii) the upper quartile = 177 cm,

    (iv) the inter-quartile range = upper quartile − lower quartile
$$= 177 - 157 = 20 \text{ cm}.$$

(c)

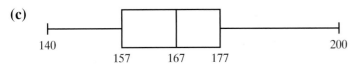

The lowest possible value that is not an outlier is the lower bound of the first interval, i.e. 140.
The highest possible value that is not an outlier is the upper bound of the last interval, i.e. 200.

(d)   (i) The 3rd percentile = 142 cm,

    (ii) the 4th decile = 163 cm.

Exercise 5K

1 The students in Year 11 at Mayfield High School are tested for their IQ.
Here are their results.

IQ	Number of students
60 to 69	3
70 to 79	8
80 to 89	14
90 to 99	43
100 to 109	47
110 to 119	28
120 to 129	4
130 to 139	3

(a) Draw a cumulative frequency polygon.

(b) Use the graph to work out estimates for
    (i) the median IQ,
    (ii) the lower quartile of these IQs,
    (iii) the upper quartile of these IQs,
    (iv) the inter-quartile range.

(c) Draw a box and whisker diagram for the data.

(d) Use your graph to estimate
    (i) the 60th percentile,
    (ii) the difference between the 2nd and 8th decile.

2 The speeds of 100 motorists were recorded one morning.
They are presented in the table below.

Speed, s (mph)	Frequency
$20 \leqslant s < 30$	5
$30 \leqslant s < 40$	13
$40 \leqslant s < 50$	18
$50 \leqslant s < 60$	32
$60 \leqslant s < 70$	27
$70 \leqslant s < 80$	3
$80 \leqslant s < 90$	2

(a) Draw a cumulative frequency polygon.
(b) Using the graph or otherwise, work out
   (i) the modal class interval,
  (ii) the median speed,
 (iii) the inter-quartile range of the speeds,
 (iv) the 82nd percentile,
  (v) the 4th and 6th deciles.
(c) Draw a box and whisker diagram.

5.11 Variance and standard deviation

The **variance** is a measure of spread, the spread about the mean of a distribution.
The variance uses *all* the data.
The square root of the variance is called the **standard deviation**.

■ **The variance is a measure of the spread of a distribution.**

For a set of numbers,

$$x_1, x_2, \ldots x_n$$

with a mean of $\bar{x}$, the variance is given by the formula

$$\text{Var}(x) = \frac{(x_1 - \bar{x})^2 + (x_2 - \bar{x})^2 + \ldots + (x_n - \bar{x})^2}{n} = \frac{\Sigma(x_i - \bar{x})^2}{n}$$

Note:
$\Sigma(x_i - \bar{x})^2$ means for each $i = 1$ to n work out $(x_i - \bar{x})^2$ and then add these together.

It can also be calculated by

$$\text{Var}(x) = \frac{x_1^2 + x_2^2 + x_3^2 + \ldots + x_n^2}{n} - \bar{x}^2 = \frac{\Sigma x_i^2}{n} - \bar{x}^2$$

■ **The positive square root of the variance is called the standard deviation. The formula for the standard deviation is**

$$\sqrt{\frac{\Sigma(x_i - \bar{x})^2}{n}} \quad \text{or} \quad \sqrt{\frac{\Sigma x_i^2}{n} - \bar{x}^2}$$

Example 24

Work out the standard deviation of the numbers

32, 34, 35, 35, 37, 37, 37, 38, 39.

The mean is

$$\bar{x} = \frac{32 + 34 + 35 + 35 + 37 + 37 + 37 + 38 + 39}{9}$$

$$= \frac{324}{9}$$

$$= 36.$$

The deviations from the mean $(x - \bar{x})$ are $-4, -2, -1, -1, 1, 1, 1, 2, 3$

So $\Sigma(x - \bar{x})^2 = 16 + 4 + 1 + 1 + 1 + 1 + 1 + 4 + 9 = 38$

and $\dfrac{\Sigma(x - \bar{x})^2}{n} = \dfrac{38}{9} = 4.222$, which is the variance.

So the standard deviation $= \sqrt{4.222}$
$$= 2.055 \text{ correct to three decimal places.}$$

Alternatively,
$$\Sigma x^2 = 32^2 + 34^2 + \dots + 39^2$$
$$= 11\,702$$
$$\frac{\Sigma x^2}{n} - \bar{x}^2 = \frac{11\,702}{9} - 36^2$$
$$= 4.222$$

Exercise 5L

1   Calculate the mean and standard deviation for the sets of data below using either of the formulae

$$SD = \sqrt{\frac{\Sigma(x_i - \bar{x})^2}{n}} \text{ or } SD = \sqrt{\frac{\Sigma x_i^2}{n} - \bar{x}^2}$$

(a) 5, 6, 10, 7, 12

(b) 8, 3, 12, 10, 7, 8, 5, 2, 5

(c) 2.1, 3.4, 6.2, 1.3, 2.9, 4.3, 5.1, 7.1, 4.2

2   Using the formula that you did *not* use in question 1, work out the standard deviation for

(a) 7, 11, 6, 8, 13

(b) 4, 9, 11, 13, 6, 8, 9, 6, 3

(c) 3.2, 2.5, 7.3, 1.4, 2.8, 4.4, 6.1, 7.3, 5.1

Standard deviation of a frequency distribution

■   **For a discrete frequency distribution the formula for the standard deviation is**

$$\sqrt{\frac{\Sigma f(x_i - \bar{x})^2}{\Sigma f}} \text{ or } \sqrt{\frac{\Sigma f x_i^2}{\Sigma f} - \bar{x}^2}$$

Example 25

Mrs Arnold gave the 30 students in her class a quick spelling test.
The marks obtained were presented in the table below.

Mark	0	1	2	3	4	5
Number of students	3	3	3	6	12	3

(a) Find the mean and standard deviation of the marks.
(b) Use the other formula to check your answer for the standard deviation.

Mark, x	f	$f \times x$	$x - \bar{x}$	$(x - \bar{x})^2$	$f \times (x - \bar{x})^2$
0	3	0	-3	9	27
1	3	3	-2	4	12
2	3	6	-1	1	3
3	6	18	0	0	0
4	12	48	1	1	12
5	3	15	2	4	12
Totals	30	90			66

(a) $\text{Mean}(\bar{x}) = \dfrac{\Sigma f x_i}{\Sigma f} = \dfrac{90}{30} = 3$

Using the first formula:

$$\text{Standard deviation} = \sqrt{\frac{\Sigma f(x_i - \bar{x})^2}{\Sigma f}} = \sqrt{\frac{66}{30}} = \sqrt{2.2} = 1.483 \text{ to 3 d.p.}$$

(b) $fx^2 = 3 \times 0^2 + 3 \times 1^2 + 3 \times 2^2 + 6 \times 3^2 + 12 \times 4^2 + 3 \times 5^2 = 336$

$$\text{SD} = \sqrt{\frac{\Sigma f x^2}{\Sigma f} - \bar{x}^2} = \sqrt{\frac{336}{30} - 9} = \sqrt{2.2} = 1.483$$

A grouped frequency distribution is treated in precisely the same
way as a discrete frequency distribution but you use the mid-interval
value for x.

Example 26

The times taken by a group of 16 students to run 100 metres are
shown below.

Time, t (seconds)	$10 < t \leqslant 11$	$11 < t \leqslant 12$	$12 < t \leqslant 13$	$13 < t \leqslant 14$
Number of students	2	5	6	3

Work out the mean and standard deviation of these data.

Replace each time interval with the mid-point of the interval.
The distribution now looks like this:

Time, t (seconds)	10.5	11.5	12.5	13.5
Number of students	2	5	6	3

The mean is $\dfrac{2 \times 10.5 + 5 \times 11.5 + 6 \times 12.5 + 3 \times 13.5}{16} = 12.125$ seconds

and the standard deviation will be

$(10.5 - 12.125)^2 = 2.640\,625$
$(11.5 - 12.125)^2 = 0.390\,625$
etc.

$$\sqrt{\frac{2 \times 2.640625 + 5 \times 0.390625 + 6 \times 0.140625 + 3 \times 1.890625}{16}} = 0.927$$

(The variance is 0.8594 correct to four places of decimals.)

Exercise 5M

1   The marks gained by a sample of 100 students in a GCSE Statistics
    examination are given in the table below.

Mark, m	$20 < m \leqslant 30$	$30 < m \leqslant 40$	$40 < m \leqslant 50$	$50 < m \leqslant 60$	$60 < m \leqslant 70$
Frequency	18	22	38	20	2

   (a) Work out an estimate of the mean mark for these 100 students.
   (b) Work out an estimate of the variance of the marks.
   (c) Work out an estimate of the standard deviation of the marks.

2   Gemma is doing a statistical project where she is examining the
    maximum speeds of a collection of cars.
    She has taken a sample of 50 cars and put them and their
    maximum speeds in a table:

Max speed (mph)	71 to 80	81 to 90	91 to 100	101 to 110	111 to 120
Number of cars	2	5	18	20	5

   Work out:
   (a) an estimate of the maximum speed of the 50 cars,
   (b) an estimate of the standard deviation of these speeds.

3   Carrie is working on a GCSE project looking at the number of
    eggs laid by different birds in their nests.
    She has the following data at hand.

Number of eggs	0	1	2	3	4	5	6
Number of nests	2	5	12	27	21	19	4

Work out:

(a) the mean number of eggs per nest,

(b) the variance of the number of eggs per nest,

(c) the standard deviation of the number of eggs per nest.

4 Calculate the mean and the standard deviation for the variable x given that:

(a) $\Sigma x^2 = 293$, $\Sigma x = 19.8$, $n = 12$

(b) $\Sigma x^2 = 3.04$, $\Sigma x = 1.26$, $n = 8$

5.12 Standardised scores

To compare values from different data sets you usually need to set up standardised scores. For this you will need to know the mean and standard deviation.

■ **Standardised score** $(z) = \dfrac{\text{score} - \text{mean}}{\text{standard deviation}}$

Example 27

Fred and Vicki did a test in English and a test in Statistics.
Both tests had a maximum mark of 100.
Their results are given in the table below.

	Fred's mark	Vicki's mark	Mean mark	Standard deviation
English	46	55	50	12
Statistics	45	42	42	8

(a) Work out Vicki and Fred's standardised scores in English and Statistics.

(b) Comment on the examination performances of the two students. Who do you think did better overall?

(a) Using

standardised score $(z) = \dfrac{\text{score} - \text{mean}}{\text{standard deviation}}$

the standardised scores for Vicki are:

English $\dfrac{55 - 50}{12} = \dfrac{5}{12} = 0.4167$ Statistics $\dfrac{42 - 42}{8} = 0$

The standardised scores for Fred are:

English $\dfrac{46 - 50}{12} = \dfrac{-4}{12} = -0.33$ Statistics $\dfrac{45 - 42}{8} = \dfrac{3}{8} = 0.375$

A negative standardised score means the result was below the mean.

(b) We can show these standardised scores in a table:

	Fred	Vicki
English	−0.33	0.4167
Statistics	0.375	0

The lowest standardised score is the −0.33 obtained by Fred in English and the best is the 0.4167 which Lisa scored in the same subject. Overall the better results appear to be those gained by Lisa, since she did not gain any standardised score which was a negative value.

Exercise 5N

1   The table shows Raji's waist measurement and his height alongside the averages of the students in his year group.

	Raji's measurement	Class average	Standard deviation
Waist (cm)	100	89	6
Height (cm)	178	165	4

Calculate the standardised score for Raji's

(a) waist measurement,

(b) height.

2   The table gives the History and Geography results for Sean, Theresa and Victoria along with the mean and standard deviation of those results for the whole year group.

	Sean	Theresa	Victoria	Year group mean	Standard deviation
History	72	58	78	61	6
Geography	34	43	51	44	5

(a) Calculate the six standardised scores.

(b) Comment on the performances of Sean, Theresa and Victoria.

Revision exercise 5

1   The marks obtained by 30 students in a test out of 50 were

32, 26, 31, 45, 28, 32, 33, 17, 24, 17
42, 32, 22, 32, 47, 24, 32, 32, 42, 48
17, 25, 21, 37, 32, 23, 40, 30, 32, 18

Work out **(a)** the mode, **(b)** the median, **(c)** the mean mark.

2 The frequency distribution gives information about the number of GCSE subjects each student is taking.

Number of subjects	6	7	8	9	10	11
Number of students	12	15	20	29	19	7

For these data work out **(a)** the mode, **(b)** the median, **(c)** the mean.

3 The ages of some people in a café are given below.

Age	Number of people
$10 \leqslant$ age < 20	7
$20 \leqslant$ age < 30	26
$30 \leqslant$ age < 40	22
$40 \leqslant$ age < 50	10

Find:
(a) the modal class interval,
(b) an estimate for the median age,
(c) an estimate for the mean age of the people in the café.

4 Work out the mean of the set of numbers
    3.02, 3.05, 3.10, 3.01 and 3.03.

5 The mid-day temperatures in Malta during one week in July are:

Mon	Tues	Wed	Thur	Fri	Sat	Sun
32	31	28	24	30	29	27

(a) Work out the mean mid-day temperature.
(b) State why you could not give the mode of the mid-day temperatures.

6 John just passed his examination, scoring 52%.
He sat three papers which were weighted 40% : 40% : 20%.
His marks were 48, 50 and x.
Work out x.

7 Work out the geometric mean of
    £12 000, £13 400, £8200, £14 500.

8 The value of a motorbike is given in this table.

Age in years	0	1	2	3	4
Value	£12 000	£10 050	£8400	£6800	£5100

Find the chain base index numbers for each year.

9   The prices of 160 second-hand cars are given below.

Price, x (£1000s)	$2 < x \leqslant 3$	$3 < x \leqslant 4$	$4 < x \leqslant 5$	$5 < x \leqslant 6$	$6 < x \leqslant 7$
Frequency	13	24	48	70	5

(a) Draw a cumulative frequency curve.
(b) Draw a cumulative frequency polygon.
(c) On each diagram mark (i) the median, (ii) the lower quartile, (iii) the upper quartile and write (iv) the inter-quartile range of the prices of these cars.
(d) Work out (i) the 35th percentile, (ii) the 8th decile of these prices.
(e) Draw a box and whisker diagram.
(f) What will be the least value of an outlier?

10 Calculate the mean and standard deviation for the variable x given that
$$\Sigma x^2 = 3000, \; \Sigma x = 240, \; n = 20.$$

11 The table shows Jennifer and Samuel's marks in the 4th Year test, along with the mean marks and standard deviations.

	Jennifer	Samuel	Mean of 4th year	Standard deviation
English	66	52	51	12
Science	43	51	40	6

(a) Calculate the four standardised scores.
(b) Comment on Jennifer's performance.
(c) Comment on Samuel's performance.

Summary of key points

1   An average is a single value used to describe a set of data.
    The mode is the value that occurs most often.
    The median is the middle number of a list of numbers after they have been put in order.
    The mean is worked out by adding up the items then dividing that total by the number of items.

2   When information is presented in a table of grouped data, the modal class interval is the class interval with the largest frequency.

3   The median can be found using the mid-point of a cumulative frequency diagram or polygon.

4 An estimated mean can be found from a grouped set of data, using the formula

$$\text{Mean} = \frac{\Sigma(f \times \text{mid-point})}{\Sigma f}$$

where Σ is 'sum of' and f is the frequency.

5 To work out the geometric mean of n numbers, multiply the numbers together and then take the nth root of the product.

$$\text{Geometric mean} = \sqrt[n]{x_1 \times x_2 \times \ldots \times x_n}$$

6 An index number shows the rate of change in quantity, value or price of an item over a period of time.

7 A price relative shows the change in price over time. It is calculated as a percentage of its value at a given base year.

8 A chain base index number tells you the annual percentage change. It is found by using the previous year as a base year and then working out the relative value of an item.

9 To calculate a weighted index number you need to:
   – calculate the index number for every element, and then
   – find the weighted average of all the elements.

10 The retail price index is a weighted mean of the price relatives of goods and services. The weightings are chosen in such a way as to show the spending habits of an average household.

11 Range = largest value − smallest value
   The lower quartile, Q_1, is the $\frac{1}{4}(n + 1)$th value
   The upper quartile, Q_3, is the $\frac{3}{4}(n + 1)$th value
   The median, Q_2, is the $\frac{1}{2}(n + 1)$th value
   The inter-quartile range = upper quartile − lower quartile

12 A set of data is divided into 100 equal parts to form percentiles. When the set of data is divided into 10 equal parts, these are called deciles.

13 A box and whisker diagram can be drawn to represent data.

14 An outlier is any value which is 1.5 (or more) times the inter-quartile range below the lower quartile or 1.5 (or more) times the inter-quartile range above the upper quartile.

15 The variance is a measure of the spread of a distribution.

16 The positive square root of the variance is called the standard deviation. The formula for the standard deviation is $\sqrt{\dfrac{\Sigma(x_i - \bar{x})^2}{n}}$

   or $\sqrt{\dfrac{\Sigma x_i^2}{n} - \bar{x}^2}$

17 For a discrete frequency distribution the formula for the standard deviation is

$$\sqrt{\frac{\Sigma f(x_i - \bar{x})^2}{\Sigma f}} \text{ or } \sqrt{\frac{\Sigma f x_i^2}{\Sigma f} - \bar{x}^2}$$

18 Standardised score $(z) = \dfrac{\text{score} - \text{mean}}{\text{standard deviation}}$

6 Scatter diagrams and correlation

6.1 Scatter diagrams

■ **You can use a scatter diagram to show whether two sets of data are related.**

Scatter diagrams should always be drawn on graph paper or by using ICT.

Example 1

A teacher thinks that students do equally well at mathematics as they do at science.

The marks out of 50 attained by 9 students in Mathematics and Science examinations are shown in the table.

Candidate	A	B	C	D	E	F	G	H	I
Mathematics mark (x)	16	20	20	28	29	31	37	39	45
Science mark (y)	15	18	22	26	29	33	38	36	45

(a) Draw a scatter diagram of these data.

(b) Explain whether the teacher's idea is correct.

(a) Plot the Science marks against the Mathematics marks on a scatter diagram like this.

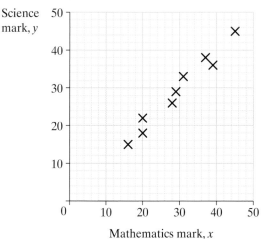

Each pair of Mathematics and related Science marks is used as coordinates to plot a point on the graph. For example, the first point plotted will have the coordinates (16, 15) (16 for Mathematics and 15 for Science). Label the axes and do not join up the points on a scatter diagram.

(b) The scatter diagram shows that as the Mathematics marks increase the Science marks also increase. There is an association between Science and Mathematics marks.

■ **If the points on a scatter diagram lie approximately on a straight line, there is a linear relationship between the two sets of data.**

Linear means straight line. This is sometimes referred to as a linear **association** between the variables.

Example 2

Ten 10 pence coins were weighed and their age, in years, noted. The results are given in the table below.

Coin	1	2	3	4	5	6	7	8	9	10
Age, x (years)	16	18	9	11	13	20	21	25	12	14
Mass, y (grams)	10.95	11.10	11.18	11.31	11.10	10.95	10.87	10.91	11.20	11.15

(a) Plot this information on a scatter diagram.

(b) Is there an association between the age and the mass of the coins? Suggest an explanation for your answer.

(a)

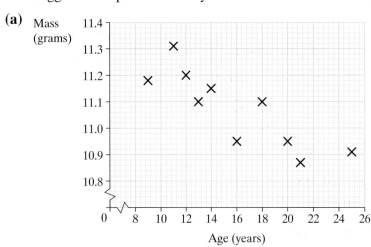

Although whenever possible you should begin your scales at zero (0), it is sometimes better to select scales that do not begin at zero. Scales should be chosen to make the plotting of all the pairs of data as easy as possible. If a scale does not begin at zero, draw attention to this fact by a break in the axis near the origin.
In this case suitable scales for age would be from 8 to 26 years, and mass could go from 10.8 to 11.4.

(b) Generally, older coins weigh less than newer ones. There is some association between the age and weight of coins. This might be because coins wear away as they are used.

Exercise 6A

1   A set of pupils sat a mock examination in English. Later they sat the final examination. The marks obtained by the candidates in both examinations were as follows:

Pupil	A	B	C	D	E	F	G	H
Mock marks, M	10	15	23	31	42	46	70	75
Final marks, F	11	16	20	27	38	50	68	70

(a) Plot the marks on a scatter diagram.

(b) Are good marks in the mock examination associated with good marks in the final examination?

2   The table below shows the marks awarded to six skaters by two judges in an ice skating competition.

Skater	1	2	3	4	5	6
Judge A	6.5	7.0	7.2	8.1	8.6	9.0
Judge B	7.4	8.2	6.4	6.8	8.5	8.5

(a) Plot a scatter diagram of these marks.

(b) Comment on how the marks of the two judges compare.

3   The table shows the number of hours of sunshine and the maximum temperature in ten British towns, on one particular day.

Town	A	B	C	D	E	F	G	H	I	J
Number of hours of sunshine	11	17	15	13	12	10	10	10	12	14
Max. temp. (°C)	13	21	20	19	15	16	12	14	14	17

(a) Using cm^2 graph paper plot this information on a scatter diagram.

(b) What does your diagram tell you about the maximum temperature as the number of hours of sunshine increases?

4   The weight and height of ten adults were as follows:

Adult	A	B	C	D	E	F	G	H	I	J
Height, H (cm)	155	163	183	198	164	178	205	203	213	208
Weight, W (kg)	58	61	85	93	70	76	84	98	100	101

It is thought that the taller a person is the more they weigh.

(a) Select suitable scales and plot a scatter diagram for these results.

(b) Comment on whether or not the height and weight are associated.

5   A general knowledge test was given to seven boys of different ages with the following results:

Boy	A	B	C	D	E	F	G
Age (years)	11.5	13	13.5	15	16	17	17
No. of correct answers	18	23	25	24	33	32	40

(a) Select suitable scales and plot a scatter diagram.

(b) Is there an association between age and the number of correct answers?

6.2 Recognising correlation

■ **Correlation is a measure of the strength of the linear association between two variables.**

The scatter diagram below shows the selling price of six second-hand bicycles in relation to their age.

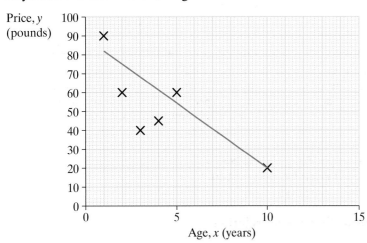

As the age of the bicycle increases the second-hand value of a bicycle decreases. This is called **negative correlation**.
Because the points are very scattered about a straight line this is called a **weak linear correlation**.

■ **Negative correlation is when one variable tends to decrease as the other increases.**

The scatter diagram below shows the petrol consumption of five cars with different engine sizes over a measured distance.

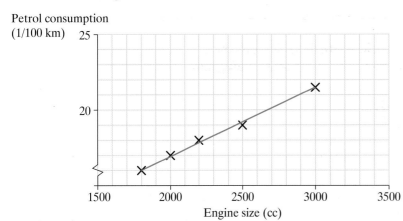

As the car engine size increases the petrol consumption increases. This is called **positive correlation**.
Because the points lie almost on a straight line this is a **strong linear correlation**.

■ **Positive correlation is when one variable tends to increase as the other increases.**

It is possible to have strong or weak positive linear correlations, and strong or weak negative correlations. The scatter diagrams below show the possibilities.

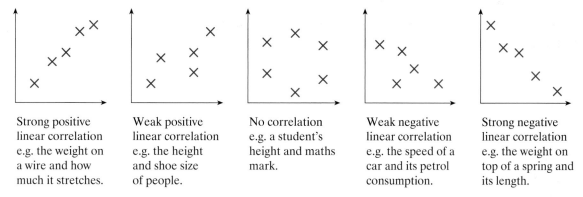

| Strong positive linear correlation e.g. the weight on a wire and how much it stretches. | Weak positive linear correlation e.g. the height and shoe size of people. | No correlation e.g. a student's height and maths mark. | Weak negative linear correlation e.g. the speed of a car and its petrol consumption. | Strong negative linear correlation e.g. the weight on top of a spring and its length. |

The middle scatter diagram shows two variables that are associated – the points lie on a circle. The linear correlation between them is, however, zero, since they do not lie close to a straight line.

■ **Association does not necessarily mean there is correlation.**

Exercise 6B

1 For the following diagrams state the type of correlation suggested.

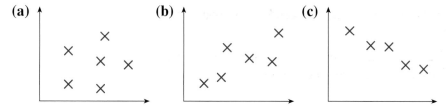

(a) (b) (c)

2 The examination grades of a sample of students in both their mock and final examinations are shown on the scatter diagram.

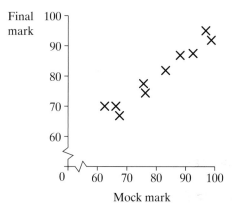

(a) Describe the correlation shown in the diagram.

(b) State the conclusion you draw regarding the relationship between the mock and the final marks of the students.

3 The number of goals scored by football teams and their positions in the league were recorded for the top twelve teams. The resulting data are shown.

Team	A	B	C	D	E	F	G	H	I	J	K	L
Goals	49	44	43	36	40	39	29	21	28	30	33	26
League position	1	2	3	4	5	6	7	8	9	10	11	12

(a) Plot a scatter diagram of these data.

(b) Describe the type of correlation.

4 It is thought that there is a correlation between the number of times a computer game is played and the score gained at the next attempt. Seven students were asked how many times they had played the game and their next score was noted with the following results:

Student	A	B	C	D	E	F	G
Number of times played, x	6	8	4	5	3	7	9
Score, y	33	37	28	30	22	37	40

(a) Draw a scatter diagram for these two variables on cm² paper.

(b) What correlation does this scatter diagram suggest?

(c) What conclusions can you draw regarding the score at the next attempt?

5 The heights and the weights of six Labrador dogs were recorded as follows:

Dog	1	2	3	4	5	6
Height, x (cm)	61	45	51	48	53	56
Weight, y (kg)	37	30	32.5	32	34	36

(a) Draw a scatter diagram of the heights and the weights of the dogs.

(b) Describe the correlation between the height and the weight of the dogs.

(c) What conclusion can you draw about the correlation between the heights and the weights of Labrador dogs?

6.3 Causal relationships

The amount of petrol a car uses depends on the size of its engine, since bigger engines use more petrol. The size of the engine *causes* the car to use more petrol. There is a **causal relationship** between the amount of petrol used and the engine size.

■ **When a change in one variable directly causes a change in another variable, there is a causal relationship between them.**

Causal relationships and correlation

With a second-hand bicycle, the older it is, the less its second-hand value. The age of the bicycle causes its value to decrease. There is a causal relationship between a bicycle's age and its second-hand price, but, because the condition of the bicycle also plays a part in its second-hand price, the correlation is not strong.

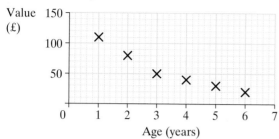

The scatter diagram below shows the number of television sets and the number of calculators sold by an electrical shop over a period of seven years.

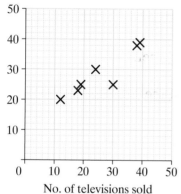

From the diagram you can see that there is a positive linear correlation between the number of televisions sold and the number of calculators sold, but there is not a causal relationship between them – buying a television does not cause you to buy a calculator. Both variables may depend on another factor, e.g. an advance in technology or an increase in wages.

■ **Correlation does not necessarily mean there is a causal relationship.**

Exercise 6C

1   Which of the following pairs of variables are likely to have a causal relationship?
    (a) A car's weight and its petrol consumption.
    (b) Sales of chocolates and sales of clothes.
    (c) Low temperature and snowfall.
    (d) Sales of computers and sales of software.

2   Dodgy Cars have eight cars for sale. The scatter diagram shows
    the ages and prices of the eight cars.

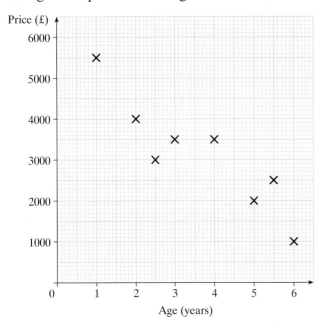

   (a) What is the price of the car that is 4 years old?
   (b) How old is the car that costs £2000?
   (c) Describe the correlation between a car's age and its price.
   (d) Is the correlation between age and price a causal relationship?

3   In a training scheme for young people, the times they took to
    reach a required standard of proficiency and their ages were as
    given in the table below:

Trainee	A	B	C	D	E	F	G	H	I	J
Age of trainee (years)	16	17	18	19	19	21	20	21	18	20
Training time (months)	8	6	9	8	12	9	10	12	7	11

   (a) Draw a grid like this on cm² graph paper.

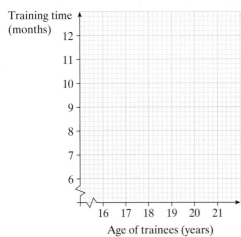

(b) Plot a scatter diagram for these data.

(c) State whether or not the diagram gives evidence of correlation, and, if it does, whether the correlation is negative or positive and if it is strong or weak.

(d) Would you say there was a causal relationship between the two variables?

4 Twelve students sat two Biology tests, one theoretical and one practical. Their marks are shown below.

Student	A	B	C	D	E	F	G	H	I	J	K	L
Marks in theory test, x	5	9	7	11	20	4	6	17	12	10	15	18
Marks in practical test, y	6	8	9	14	21	8	7	16	15	8	18	18

(a) Draw a scatter diagram to represent these data.

(b) Describe the correlation between the theory and the practical tests.

(c) Is there a causal relationship between the two variables?

5 The table shows the number of cars per 100 of the population and the number of road deaths per 100 000 of the population for ten different countries.

Country	Cars per 100 of population, x	Road deaths per 100 000 of population, y
A	31	15
B	32	35
C	55	23
D	64	20
E	28	26
F	19	19
G	34	22
H	38	22
I	48	31
J	60	36

(a) Draw a scatter diagram to represent these data.

(b) Describe the correlation between the number of cars and the number of accidents.

(c) Is there a causal relationship between these variables?

6.4 Drawing lines of best fit on scatter diagrams

Points on a scatter diagram are strongly correlated if they lie almost along a straight line.

The scatter diagram shows the heights and shoe sizes of ten pupils from a class of thirty.

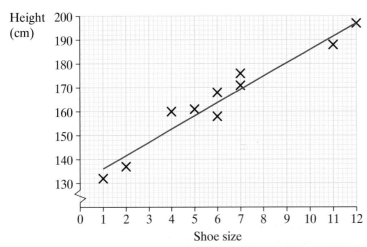

Because height and shoe size are strongly correlated, a straight line can be drawn that passes as close to (or through) as many points as possible. This line is called the **line of best fit**.

The line of best fit has been drawn by eye.

Example 3

The table shows the height above sea level, x (metres), and the temperature, y (°C), on the same day at nine different places.

Place	A	B	C	D	E	F	G	H	I
Height, x (m)	1400	400	280	800	920	600	560	1220	680
Temp, y (°C)	7	15	19	9	10	14	12	8	14

(a) Draw a scatter diagram and add a line of best fit.

(b) Use your line of best fit to estimate the air temperature at 600 m above sea level.

(a)

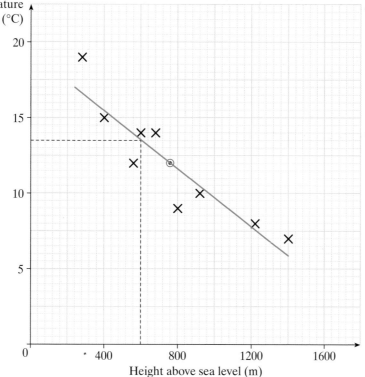

Air temperature (°C) plotted against Height above sea level (m)

The line of best fit does not necessarily pass through any of the points on the scatter diagram.

The line of best fit is drawn through the mean of both sets of data $(\bar{x}, \bar{y})$.

(b) 13.5 °C

■ **A line of best fit is a straight line drawn so that the plotted points on a scatter diagram are evenly scattered either side of the line.**

In Example 3 the air temperature at 600 m above sea level was estimated from the line of best fit. The line of best fit is a model for how air temperature changes with height above sea level.

■ **A line of best fit is a model for the association between the two variables.**

The mean point

To help with fitting a line of best fit you can calculate the means of the two variables and plot them as a point on the graph.

In an examination you will be told if you are required to work out and use the mean point or just draw the line by eye.

■ **A line of best fit should pass through the mean point $(\bar{x}, \bar{y})$.**

For Example 3,

$$\text{Mean height} = \frac{1400 + 400 + 280 + 800 + 920 + 600 + 560 + 1220 + 680}{9}$$

$$= 762.2$$

Mean temperature $= \dfrac{7 + 15 + 19 + 9 + 10 + 14 + 12 + 8 + 14}{9}$

$\qquad\qquad\qquad = 12$

So the mean point $(\bar{x}, \bar{y}) = (762.2, 12)$. This point is circled on the scatter diagram, and the line of best fit passes through this point.

Exercise 6D

1   An investigation was undertaken into the length of main road, x (in 1 000 000 miles), and the number of road accident injuries per year, y (in 100 000), for seven industrialised countries. The results are shown in the table.

Country	A	B	C	D	E	F	G
Main road length, x	12.4	28.5	31.2	45.8	18.4	44.4	18.4
Injuries, y	3.1	2.4	4.5	2.2	1.7	1.1	2.7

(a) Draw a scatter diagram for these data.
(b) Would you draw a line of best fit for these data? Give reasons for your answer.

2   The following table shows the examination marks of eight students in English and French.

Student	A	B	C	D	E	F	G	H
English, x	10	20	30	32	49	52	61	74
French, y	20	24	35	30	48	59	72	80

(a) Find the mean of each set of marks.
(b) Draw a scatter diagram for your data. Both axes can start at 0.
(c) Plot the mean point found in (a) and draw a line of best fit through it.

3   (a) Draw a horizontal axis from 20 to 40 and a vertical axis from 0 to 12.
(b) Plot the following points on your diagram: (22, 3), (24, 4), (25, 4), (29, 7), (32, 8), (36, 10).
(c) Find the mean point and mark it on your diagram.
(d) Draw a line of best fit.

4   The following table shows the heights and the weights of ten boys.

Boy	A	B	C	D	E	F	G	H	I	J
Height, x (cm)	130	129	133	135	136	140	142	145	150	160
Weight, y (kg)	30	33	33	38	37	40	44	52	61	72

(a) Draw a scatter diagram of these data.

(b) Find the mean height and the mean weight and mark the mean point on your diagram.

(c) Draw in a line of best fit.

(d) What sort of correlation does your diagram show?

6.5 Using scatter diagrams and lines of best fit

Lines of best fit can be used to estimate other values from the graph.

In Example 3 the line of best fit was used to estimate the air temperature at 600 m above sea level. The readings were for heights from 400 m to 1220 m, so 600 m fell within this range.

In order to extrapolate you must extend outside the range of given values. You should bear this in mind when selecting scales for your scatter diagram.

■ **Interpolation is when you find values within the range of values you are given.**

■ **Extrapolation is when you find values outside the range of values you are given.**

Example 4

The table shows the result of an experiment on the effect of water temperature on the number of heartbeats per minute of *Daphnia* (a water flea).

Observation	1	2	3	4	5	6
Water temperature, x (°C)	5	10	15	20	25	30
Number of heartbeats/min	110	200	240	300	380	400

(a) Draw a scatter diagram of these data and add a line of best fit that passes through the mean point.

(b) Using the line of best fit:
   (i) interpolate to estimate the number of heartbeats per minute when the water temperature is 22 °C
   (ii) extrapolate to estimate the number of heartbeats per minute when the water temperature is 35 °C.

(a) Mean temperature $= \dfrac{5 + 10 + 15 + 20 + 25 + 30}{6} = 17.5$

Mean heartbeat $= \dfrac{110 + 200 + 240 + 300 + 380 + 400}{6} = 271.7$

The mean point is (17.5, 271.7).

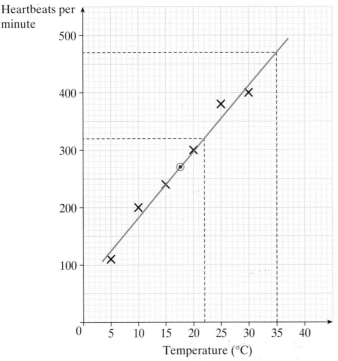

**Scatter diagram showing number of
heartbeats at different temperatures**

(b) From the diagram:
  (i) number of heartbeats when the temperature is 22 °C = 320
  (ii) number of heartbeats when the temperature is 35 °C = 470

Interpolated values are reasonably accurate.
You must be careful with extrapolated values – the further the
extrapolation is outside the range of given values, the less
reliable the estimated value.

The dangers of
extrapolation are shown by
this example. The estimate
for the number of
heartbeats at 35 °C was 470,
the actual value was
observed to be 389. An
estimate at 40 °C would be
520: in fact the correct
value is 0 since the water
flea would be dead in water
at this temperature.

Exercise 6E

1 A measure of personal fitness is the time taken for a person's
pulse rate to reach normal after strenuous exercise. Gordon
recorded his pulse rate y at time x minutes after finishing some
strenuous exercise. The results are shown in the table.

Time, x (min)	0.5	1.0	1.5	2.0	2.5	3.0	3.5
Pulse rate, y (beats/min)	125	113	110	94	81	83	71

(a) Draw a scatter diagram and add a line of best fit.
(b) Estimate Gordon's pulse rate 2.5 minutes after stopping
exercise.
(c) Estimate Gordon's pulse rate 5 minutes after stopping
exercise.
(d) Which of the two estimates is the most reliable?

2 A bar was supported at its ends in a horizontal position and various weights, x (kg), were hung from the mid-point of the bar. The deflection (how much the middle of the bar sags), y (cm), was recorded each time. The results are shown in the table.

Mass, x (kg)	20	25	30	35	40	45	50
Deflection, y (cm)	0.20	0.32	0.34	0.40	0.49	0.59	0.65

(a) Draw a scatter diagram for these data.

(b) Draw a line of best fit on your scatter diagram.

(c) Estimate what the deflection would be under a weight of 28 kg.

(d) Estimate what the deflection would be under weights of 15 kg and 55 kg.

(e) Which of your three estimates is likely to be the most accurate?
Explain your answer.

3 Ten students were selected at random from those visiting the tuck shop at mid-morning break. The students were asked their age and how much pocket money they got each week. The results are shown in the table.

Student	1	2	3	4	5	6	7	8	9	10
Age, x (years)	17	16	18	13	10	$11\frac{1}{2}$	14	11	15	12
Pocket money, y (£)	12	15	20	10	2	2.25	10.5	2.5	11	13

(a) Draw a scatter diagram of these data.

(b) Add a line of best fit and use this to predict how much a child of $13\frac{1}{2}$ should get.

(c) Explain why you would not bother to extrapolate to find how much a 25-year-old would get for pocket money.

4 The sales in units of £1000 of a certain company for the years 1996 to 2002 are given in the table.

Year, x	1996	1997	1998	1999	2000	2001	2002
Sales, y (£1000s)	65	70	73	78	83	83	88

(a) Draw a scatter diagram for these data.

(b) Add a line of best fit that passes through the mean point, and use your line to predict the sales for 2004. Comment on the validity of this estimate.

5   The length, y mm, of a metal rod was measured at various temperatures, x °C, giving the following results.

Temperature, x (°C)	60	65	70	75	80	85
Length, y (mm)	100.2	100.8	101.8	102	103.4	104.5

(a) Draw a scatter diagram for these data, and add a line of best fit that passes through the mean point.

(b) Use your line to predict the length of the rod when the temperature was 68 °C and what it would be at 100 °C. Comment on the validity of these estimates.

6   The percentage of 'A'-grades awarded each year in a particular examination are shown in the table.

Year	1994	1995	1996	1997	1998	1999	2000	2001
% 'A'-grades	14.7	15.4	15.7	15.7	16.8	17.5	17.8	18.6

(a) Draw a scatter diagram for these data, and comment on the correlation it shows.

(b) Draw a line of best fit through the mean point of these data, and use it to predict the percentage of 'A'-grades in the 2002 examination.

6.6 Finding the equation of a line of best fit

You should be familiar with the equation for a straight line in the form $y = mx + c$. In statistics it is more usual to use it in the form $y = ax + b$.

The equation of the line $y = ax + b$ has a gradient a, and its intercept on the y-axis is $(0, b)$.

Example 5

The graph shows the length of a spring, y (in cm) under different loads, x (in kg).
Find the equation of the line.

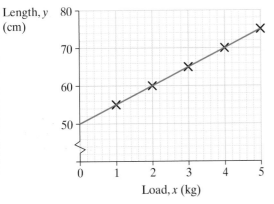

The intercept on the y-axis is the point $(0, 50)$ so $b = 50$.
When x increases from 1 to 4, y increases from 55 to 70. When x increases by 3, y increases by 15.

For every increase of 1 in x, y increases by $\dfrac{15}{3} = 5$.

So the gradient is 5 and $a = 5$.

If the graph had not shown the intercept on the y-axis, then you could have found b another way. Since every point on the graph fits the law $y = ax + b$ and $(1, 55)$ is a point on the line

$$55 = (5 \times 1) + b \text{ and } b = 55 - (5 \times 1) = 50.$$

Thus the equation is $y = 5x + 50$

> You could use any two points on the line.

> You can use any point on the line.

Finding the values of *a* and *b*

1 Select two points on the graph. These should be well apart and at points where the values of x and y are whole numbers if possible. Let the point with the lower x value be (x_1, y_1) and let the point with the higher x value be (x_2, y_2).

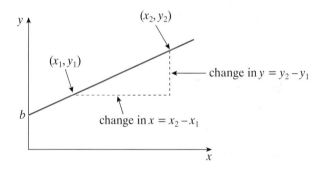

Then $a = \text{gradient} = \dfrac{\text{change in } y}{\text{change in } x} = \dfrac{y_2 - y_1}{x_2 - x_1}$.

2 If the x values on the scatter diagram go down to 0, then the value of b can be read directly from the graph.

If the x values do not go down to 0, the formula can be rearranged to give $b = y - ax$, and by substituting in this formula the values from one of the points on the line and the value for a, the value of b may be calculated.

e.g. $b = y_1 - ax_1$ or $b = y_2 - ax_2$

■ **The equation of a line of best fit is $y = ax + b$, where a is the gradient of the line and b is the intercept with the y-axis.**

$$a = \frac{y_2 - y_1}{x_2 - x_1} \text{ and } b = y_1 - ax_1 \text{ or } b = y_2 - ax_2$$

> You may already have come across this equation in the form $y = mx + c$.

Example 6

Find the equation of the line of best fit for Example 3.

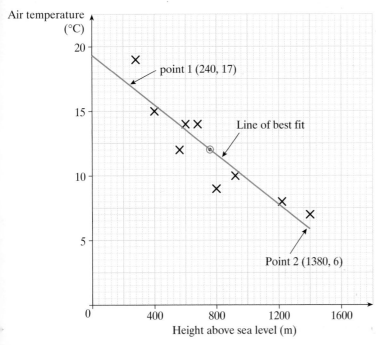

Two points on the line are $(x_1, y_1) = (240, 17)$ and $(x_2, y_2) = (1380, 6)$.

$$a = \frac{y_2 - y_1}{x_2 - x_1} = \frac{6 - 17}{1380 - 240} = -0.009\,649\,1$$

From the graph, $b = 19.3$

The gradient may look larger on the graph but the two scales are very different.

(In this case the value of b has been read from where the line cuts the y-axis on the scatter diagram. You cannot read the value exactly but you can make a good estimate.)

Alternatively, $b = y_2 - ax_2$
$$= 6 - (-0.009\,649\,1) \times 1380$$
$$= 6 + 13.316$$
$$= 19.316 \text{ (by calculation)}$$

The equation of the line of best fit is $y = -0.0096x + 19.3$

Exercise 6F

1   The scatter diagram shows the results of a survey on the thickness of the soles of trainers, y, and the number of years that they had been worn, x.

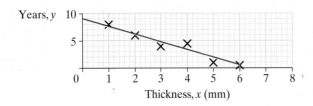

For the line of best fit shown in the diagram, find:

(a) the gradient of the line

(b) the intercept of the line with the y-axis

(c) the equation of the line of best fit.

2 Work out the equation of the line of best fit for the scatter diagram shown below.

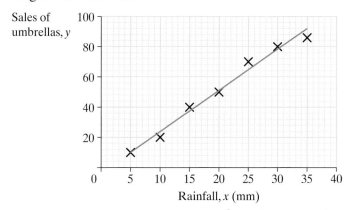

3 The scatter diagram shows the coursework and exam marks for seven students. Copy the diagram.

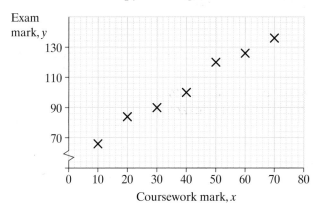

Draw in a line of best fit and work out the equation of the line of best fit that passes through the mean point.

4 Ten students were selected at random from those visiting the tuck shop at mid-morning break. The students were asked their age and how many hours they watched television each week. The results are shown in the table.

Student	1	2	3	4	5	6	7	8	9	10
Age, x (years)	17	16	18	13	10	$11\frac{1}{2}$	14	11	15	12
Hours of TV watching, y	12	15	20	10	2	2.25	10.5	2.5	11	13

(a) Draw a scatter diagram of these data.

(b) Add a line of best fit that passes through the mean point. Calculate the equation of the line of best fit and use it to predict how many hours a child of $16\frac{1}{2}$ watches television.

(c) Explain why you would not bother to extrapolate to find how many hours a 40-year-old would watch television.

5 (a) For the data in Exercise 6E question 4, find the equation of the line of best fit.

(b) Use the equation to predict the sales for the year 2004. Comment on the validity of this estimate.

6 For the data in Exercise 6E question 5:

(a) Calculate the equation for your line of best fit in the form $y = ax + b$, and use it to predict the length of the rod at:

  (i) 68 °C.

  (ii) 100 °C.

  Comment on the validity of these estimates.

(b) What does the constant b represent and what does the constant a represent in this case?

7 For the data in Exercise 6E question 6:

Use 1994 = Year 1

(a) Find the equation for the line of best fit. Use this equation to predict the percentage of 'A'-grades in the 2002 examination.

(b) Explain why this equation would not be reliable for forecasting the year when the percentage of 'A'-grades reaches 50%.

8 The manager of a factory decided to give the workers an incentive by introducing a bonus scheme. After the scheme was introduced the manager thought that the workers might be making more faulty products because they were rushing to make articles quickly. A study of the number of articles rejected, y, and the amount of bonus earned, £x, gave the figures shown in the table.

Employee	A	B	C	D	E	F	G	H
Bonus, x (£)	14	23	17	32	16	19	18	22
Number of rejects, y	6	14	5	16	7	12	10	14

(a) Draw a scatter diagram for these data and add a line of best fit that passes through the mean point.

(b) What sort of correlation is there between the two variables? What does this mean in terms of the manager's belief?

(c) Calculate the equation of the line of best fit.

(d) If the maximum number of rejects acceptable is 9, what level should the maximum bonus be set at?

9 The height of a seedling, y millimetres, x weeks after it is planted is given in the table.

x	5	6	7	8	9	10
y	102	111	123	135	148	153

(a) Plot a scatter diagram for these data.

(b) Draw a line of best fit on your scatter diagram that passes through the mean point.

(c) Calculate the equation of the line of best fit in the form $y = ax + b$.

(d) What do the constants a and b represent in this case?

(e) Use the equation found in **(c)** to estimate the height of the seedling after 20 weeks. How accurate will this estimate be?

6.7 Using ICT to plot scatter diagrams and lines of best fit

Example 7

The ages in months and the weights in kilograms of a random sample of nine babies are shown.

Baby	A	B	C	D	E	F	G	H	I
Age, x (months)	1	2	2	3	3	3	4	4	5
Weight, y (kg)	4.4	5.2	5.8	6.4	6.7	7.2	7.6	7.9	8.4

(a) Draw a scatter diagram for these data.

(b) Draw a line of best fit.

(a) From Microsoft Word click on the **Insert Microsoft Excel worksheet** button on the tool bar.

- Enter the information on a spreadsheet as shown.

- Select cells A1 to B10 and use the **chart wizard** button on the toolbar.

- (Step 1 of 4) From **chart type** select **XY (Scatter)** then select the top left diagram. Click **Next**.

- (Step 2 of 4) Click **Next**.

- (Step 3 of 4) Change title to 'Weights of babies'.
  Value X-axis 'Age (months)'
  Value Y-axis 'Weight (kg)'
  Click on **Gridlines** at the top then click to get ticks in all four squares.
  Click **Next**.

	A	B
1	Age	Weight
2	1	4.4
3	2	5.2
4	2	5.8
5	3	6.4
6	3	6.7
7	3	7.2
8	4	7.6
9	4	7.9
10	5	8.4

- (Step 4 of 4). Click on **Finished**.
  Scatter diagram appears.

- To change the X-axis scale, move pointer to X-axis so 'Value
  X-axis' appears and click.
  Click on **Format** on the toolbar, then **Selected axis** on the
  drop down menu.
  Click on **Scale**. Change the Maximum to 6, Major unit to 1,
  and Minor unit to 0.2. (Y-axis can be changed in the same
  way if needed.)

- The scatter diagram can now be adjusted for size. It looks
  like this:

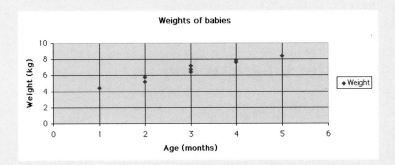

Click on **Age (month)**, then **Format** on the tool bar, then
Selected axis title on drop down menu. Click on **Font**,
change size to 12. Repeat with other 'Weight (kg)' and title.

To add a line of best fit

Click on **Chart** on the toolbar, then **Add trend line**. Select the
top left diagram, the linear trend line, and click on **OK**.

To calculate the equation of a line of best fit using a spreadsheet

When data are entered on the spreadsheet do the following.

- In A11 on the spreadsheet enter the word 'slope'.

- In A12 enter '=SLOPE(B2:B10,A2:A10)' and click on the
  tick. Value of a appears in A12.

- In B11 enter the word 'Intercept'.

- In B12 enter '=INTERCEPT(B2:B10,A2:A10)' and click on
  the tick. The value of b appears in B12.

(Actual values $a = 1.041\,67$, $b = 3.497\,22$)

If you are drawing the trend
line on a chart you can get
the equation calculated and
printed on your chart by:
- Click on **Chart** on the
  toolbar.
- Click on **Add trend line**.
- Select **Options** and point
  to the square for **Display
  equation on chart**.
Line and equation should
both appear.

6.8 Fitting a line of best fit to a non-linear model of the form $y = ax^n + b$ and $y = ka^x$

Example 8

In a biological investigation the lengths and areas of nine privet leaves were as shown.

Length, x (mm)	8	12	18	20	26	33	38	40	45
Area, y (mm²)	25	48	124	162	220	315	420	515	650

The scatter diagram for these two variables is shown below.

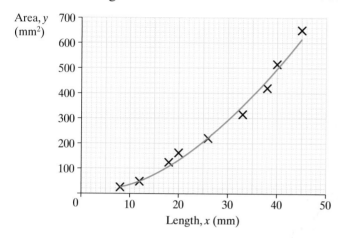

In this case a straight line is not the best model for the association between the two variables. A curve is a better model. We call this fitting a **non-linear model**.
This curve is of the form $y = ax^n + b$.

In an examination you will be told the form of curve to use for your model.

The appearance of curves of the form $y = ax^n + b$ varies according to the value of n. You will need to be able to recognise and use values for n of 2, -1 and $\frac{1}{2}$.

These give the equations

$$y = ax^2 + b, \; y = ax^{-1} + b = \frac{a}{x} + b \text{ and } y = ax^{\frac{1}{2}} + b = a\sqrt{x} + b$$

Their graphs look like this:

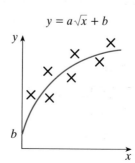

Example 9

The data in the table are thought to follow the equation $y = \dfrac{a}{x} + b$.

x	1	2	3	4	5
y	5	4.1	3.6	3.5	3.2

(a) Plot these data.

(b) Draw a straight line of best fit, and a curve of the form $y = \dfrac{a}{x} + b$.

(c) Comment on the fit of these lines.

(a) (b)

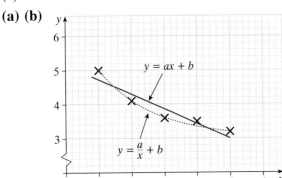

(c) The line of the form $y = \dfrac{a}{x} + b$ fits better.

Fitting a line of best fit to a non-linear model of the form $y = ka^x$

Sometimes a curve of the form $y = ka^x$ may be a better shape than the type shown above.

The function $y = ka^x$ where a is a positive number and x is a variable, is called an **exponential function**. An exponential function models the way things grow and the way things decay.

Example 10

The graph below shows the value, v, of an investment after x years. The initial investment is £1000 and the interest rate is 10%.

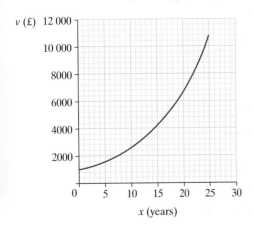

> You will be told which model, $y = ax^n + b$ or $y = ka^x$, you should consider.

This shows exponential growth and has the law $y = ka^x (a > 1)$.

The chart below shows the value of an investment of £1000 if you spent 10% of the remaining sum every year.

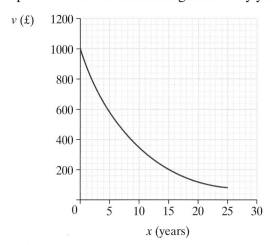

This graph shows exponential decay and has the law $y = ka^x (a < 1)$.

Exercise 6G

1   Suggest an equation for each of the following diagrams.

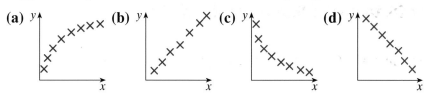

2   A scatter diagram of the following data suggests that an equation of the form $y = ax^2 + b$ would be a suitable model.

Maximum diameter of apples, x (cm)	1.1	1.7	2.2	2.4	3.3	4.0	5.2	6.4
Surface area, y (cm²)	5	10	17	20	32	53	85	136

   (a) Plot the results on a scatter diagram.
   (b) Draw a line of best fit through the points, and a curve of the appropriate shape.
   (c) Which line fits best?

3   The cost per component, y, of setting up tooling for the components' manufacture, and the number of components produced, x, are shown in the table below.

Number of components, x	1	2	3	4	5
Cost/component, y (£)	5	4.1	3.6	3.5	3.2

The data in the table are thought to follow the equation $y = \dfrac{a}{x} + b$.

(a) Draw a scatter diagram.

(b) Draw a line of best fit and a curve of the suggested shape.

(c) State which of the two models is most suitable.

4   In an experiment with a simple pendulum, a large weight was fixed at the end of a thin piece of wire. The length, x (cm), of the wire and the time for one oscillation of the pendulum, y (seconds), were recorded. The results are given in the table.

Length, x (cm)	10	20	30	40	50
Time, y (seconds)	0.6	0.9	1.1	1.3	1.4

(a) Draw a scatter diagram for these data.

(b) Which of the following equations might be a suitable model?

    (a) $y = a\sqrt{x} + b$ (b) $y = ax^2 + b$

5   The current, y, flowing through two parallel resistors is measured as the resistance, x, of one of the resistors is varied. The result is shown in the table.

Resistance, x (ohm)	1	2	3	4	5
Current, y (amp)	5	4.1	3.5	3.5	3.4

(a) Plot these points on a scatter diagram.

(b) Decide if a curve or a straight line would be the best model.

(c) Write down the general equations of two possible curves that might model the association between x and y.

6.9 Spearman's rank correlation coefficient r_s

Correlation is a measure of the strength of the linear association between two variables.

Spearman's rank correlation coefficient is a *numerical* measure of the correlation between two sets of data. It tells us how close the agreement is.

The basic idea behind this coefficient is **ranking**.

Example 11

Two judges were judging a gardening competition. There were six entries in the sweet pea section. The marks out of 12 awarded by the two judges x and y are shown in the table. Rank these data.

Entry	A	B	C	D	E	F
x value	1	2	5	8	6	3
y value	6	3	8	11	7	4

To rank the observations, you assign:
the largest value the rank 1
the next highest the rank 2
the third highest the rank 3
… and so on until all are ranked.

You could equally well rank the smallest value at 1, the next smallest at 2, etc. This makes no difference to the answer, provided you do the same for both variables.

Ranking can be done in the form of a table:

x value	*y* value	*x* rank	*y* rank
1	6	6	4
2	3	5	6
5	8	3	2
8	11	1	1
6	7	2	3
3	4	4	5

To use Spearman's rank correlation coefficient both variables must be ranked.

Spearman's rank correlation coefficient is concerned with the difference, d, in the ranking of the two sets of data.

d = rank of the x observation − rank of the y observation.

■ **Spearman's rank correlation coefficient (r_s) is given by**

$$r_s = 1 - \frac{6\Sigma d^2}{n(n^2 - 1)}$$

where n = number of sets of readings.

The value of r_s varies from −1 to 1.

This formula is given on the formula sheet.
Remember that Σd^2 means the sum of all the differences squared.

Example 12

Work out Spearman's rank correlation coefficient between the two sets of marks in Example 11.

x value	*y* value	*x* rank	*y* rank	*d*	*d²*
1	6	6	4	+2	4
2	3	5	6	−1	1
5	8	3	2	+1	1
8	11	1	1	0	0
6	7	2	3	−1	1
3	4	4	5	−1	1
				$\Sigma d^2 = 8$	

$$r_s = 1 - \frac{6\Sigma d^2}{n(n^2 - 1)}.$$

$$= 1 - \frac{6 \times 8}{6(36 - 1)}$$

$$= 1 - 0.2286$$

$$= 0.7714$$

Spearman's rank correlation coefficient is 0.7714.

Exercise 6H

1   The following marks were given to 11 candidates taking a driving test examination. Rank the numbers.

    49, 44, 43, 36, 40, 39, 29, 28, 30, 33, 26

2   A factory gave eight of their apprentices a practical test before and after a short course. The marks gained before the course, y, and those gained after the course, x, are given in the table. Separately rank each of the sets of marks.

Marks after, x	12	22	40	33	18	25	14	4
Marks before, y	10	30	45	12	28	18	19	4

3   In an investigation into smoking, eight patients were assessed for the amount of lung damage they had suffered, x, and were asked how many years they had been smoking, y. Replace each of the numbers in the table below by their separate rank.

Lung damage, x	42	56	12	6	83	44	58	23
Years smoked, y	16	18	14	19	22	13	17	25

4   Calculate Spearman's rank correlation coefficient for each of the following.
    (a) $n = 12$, $\Sigma d^2 = 18$
    (b) $n = 6$, $\Sigma d^2 = 14$
    (c) $n = 8$, $\Sigma d^2 = 100$

5   Apples were ranked for flavour (rank 1) and juiciness (rank 2). The ranks are shown in the table. Copy and complete the table and calculate r_s.

Rank 1	Rank 2	d	d^2
1	5		
2	4		
3	1		
4	6		
5	3		
6	2		

6 The following table shows the rank orders put on six collie dogs by two judges.

Collie	A	B	C	D	E	F
Judge 1	3	1	6	2	5	4
Judge 2	2	3	6	1	4	5

Work out Spearman's rank correlation coefficient.

7 The table shows the marks given by two judges to six skaters in an ice skating competition.

Skater	A	B	C	D	E	F
Judge 1	6.5	8.2	9	6	8	7.5
Judge 2	7	8.4	8.6	6	9	6.8

(a) Rank the marks of each judge.

(b) Work out Spearman's rank correlation coefficient.

6.10 Interpretation of Spearman's rank correlation coefficient

Consider the two extreme cases: when the rankings are in complete agreement and when one set of rankings is the reverse of the other.

x rank	y rank	d	d^2
1	1	0	0
2	2	0	0
3	3	0	0
4	4	0	0
5	5	0	0
6	6	0	0
			$\Sigma d^2 = 0$

x rank	y rank	d	d^2
1	6	-5	25
2	5	-3	9
3	4	-1	1
4	3	1	1
5	2	3	0
6	1	5	25
			$\Sigma d^2 = 70$

$$r_s = 1 - \frac{6\Sigma d^2}{n(n^2 - 1)} \qquad\qquad r_s = 1 - \frac{6\Sigma d^2}{n(n^2 - 1)}$$

$$= 1 - \frac{6 \times 0}{6(36 - 1)} \qquad\qquad r_s = 1 - \frac{6 \times 70}{6(36 - 1)}$$

$$= 1 \qquad\qquad\qquad\qquad = 1 - 2$$

$$\qquad\qquad\qquad\qquad\qquad = -1$$

The scatter diagrams for each of these are shown.

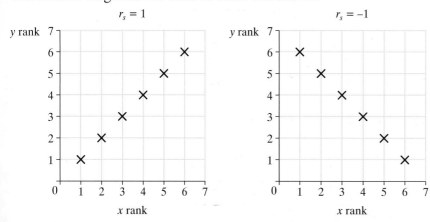

Thus if $r_s = 1$ there is perfect positive correlation, that is to say there is complete agreement between the two rankings.

If $r_s = -1$ there is perfect negative correlation: there is complete disagreement between the two rankings.

If $r_s = 0$ there is no correlation: there is no agreement or disagreement between the rankings.

The value of Spearman's rank correlation coefficient r_s gives a measure of how well the ranks agree.

- **If r_s is close to 1 there is a strong positive linear correlation, and there is close agreement between the rankings.**

- **If r_s is close to -1 there is a strong negative linear correlation, and there is close to complete disagreement between the rankings.**

- **If r_s is close to zero then there is no linear correlation, and so no agreement or disagreement between the rankings.**

You can get shades of correlation in between as shown.

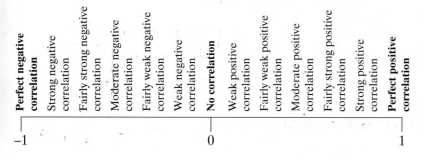

In Example 12 where $r_s = 0.7714$ you would describe it as a fairly strong positive linear correlation of the ranks.

Example 13

Two experts tried to rank eight antiques in order of value. The results are shown in the table.

Antique	1	2	3	4	5	6	7	8
Rank of judge A	8	2	7	1	3	4	6	5
Rank of judge B	8	2	6	1	4	3	7	5

Calculate Spearman's rank correlation coefficient for these data and comment on the result.

In this case the ranking has already been done, so all that is needed is to add two rows to the table for d and d^2.

Antique	1	2	3	4	5	6	7	8
Rank of judge A	8	2	7	1	3	4	6	5
Rank of judge B	8	2	6	1	4	3	7	5
d	0	0	1	0	−1	1	−1	0
d^2	0	0	1	0	1	1	1	0

$\Sigma d^2 = 4$

$$r_s = 1 - \frac{6\Sigma d^2}{n(n^2 - 1)}$$
$$= 1 - \frac{6 \times 4}{8(8^2 - 1)}$$
$$= 1 - 0.0476$$
$$= 0.9524$$

There is a strong positive correlation between the rankings of the two judges. The two judges are in close agreement.

Exercise 6I

1 Two teachers were asked to rank 10 students in a fancy dress competition. The ranks they gave are shown in the table.

Child	A	B	C	D	E	F	G	H	I	J
Teacher 1	1	9	3	6	8	2	7	4	10	5
Teacher 2	8	4	7	1	3	5	6	9	10	2

Calculate a value for Spearman's rank correlation coefficient, and comment on its value.

2 The eight photographers who entered a photographic competition were ranked by each of two judges. The Spearman's rank correlation coefficient was calculated for the rankings, the resulting value being -0.92. Comment on the agreement between the two judges.

3 In a song contest six songs were given marks out of 10 by two judges. The marks awarded are shown in the table.

Song	1	2	3	4	5	6
Judge A	3	8	9	2	7	5
Judge B	2	6	9	3	8	7

(a) Rank the two sets of marks and calculate Spearman's rank correlation coefficient.

(b) Comment on the degree of agreement between the two judges.

4 Students in Year 11 at a school can choose to study from a selection of ten subjects. The numbers of boys and girls in each of the subject classes are shown.

Subject	Art	Bio	Chem	Eco	Eng	French	Geog	Hist	Math	Phy
Girls	17	18	10	21	22	25	23	19	20	2
Boys	15	10	22	20	42	19	16	14	17	12

Use Spearman's rank correlation coefficient to see whether or not there is any agreement between the girls' choices and the boys' choices of subjects.

Remember to rank first.

5 In a cookery competition two tasters gave each of eight dishes a mark out of 50. The results are shown.

Dish	A	B	C	D	E	F	G	H
Taster 1	29	35	40	38	34	47	28	36
Taster 2	29	35	45	38	26	40	28	30

(a) Calculate Spearman's rank correlation coefficient.

(b) How would you describe the agreement between the two judges?

Remember to rank first.

6 Gemma believes that people with long surnames will have been given short first names. She decides to see if this idea is true and counts the number of letters in the first names and in the surnames of ten children chosen at random from the register. The table shows the results.

Person	1	2	3	4	5	6	7	8	9	10
Length of first name	5	9	8	7	11	12	6	10	4	3
Length of surname	5	11	12	4	14	7	3	9	6	10

(a) Calculate Spearman's rank correlation coefficient for these data.

(b) How true is Gemma's belief?

7 The table shows the initial weights, x (g), of eight insects just after hatching and their weights, y (g), after feeding for 25 days.

Insect	1	2	3	4	5	6	7	8
Initial weight, x (g)	0.74	0.77	0.79	0.83	0.9	0.96	0.98	1.10
Final weight, y (g)	0.85	0.90	0.95	1.00	0.98	1.10	1.07	1.30

(a) Calculate Spearman's rank correlation coefficient for these data.

(b) What can you conclude about the final weight in relation to the original weight?

Tied ranks

You may find two or more observations that are equal in value. These are called **tied** values. When two or more values are tied they are each given the mean value of the ranks that they would have had if they were not tied.

Consider the numbers 93, 87, 74, 74, 72, 60, 45.

These two values are tied.

The two numbers 74 would have been ranked 3 and 4 so they are given the rank $\dfrac{3+4}{2} = 3.5$.

The numbers and their ranks are shown in the table:

Number	93	87	74	74	72	60	45
Rank	1	2	3.5	3.5	5	6	7

> You will not be set examination questions that contain tied ranks. But you may need to use them in your project.

> If there are many sets of tied ranks the value given by the Spearman formula will not be very reliable.

Exercise 6J

Questions on tied ranks such as these will not be set in an examination. Tied ranks may occur in your coursework.

1 Two people are asked to give marks out of 10 to six different services provided by the local council. The marks are given in the table below.

Service	A	B	C	D	E	F
Person 1	5	4	7	3	9	5
Person 2	6	3	8	3	10	5

(a) Rank these data.

(b) Calculate Spearman's rank correlation coefficient for the ranks.

(c) Comment on the agreement between the two people's opinions.

2 Each of seven teams in a school hockey league had goals scored for and against them as shown in the table.

Team	A	B	C	D	E	F	G
Goals for	39	40	28	28	26	30	42
Goals against	22	22	27	42	24	38	23

(a) Calculate Spearman's rank correlation coefficient for these data.

(b) Interpret your result.

3 A teacher wrote down the order in which he thought six students would finish in their statistics test, before giving them the test. The results are shown in the table.

Student	A	B	C	D	E	F
Order	1	=2	3	=2	5	4
Statistics mark	84	68	37	60	37	20

(a) Rank these data.

(b) Calculate Spearman's rank correlation coefficient.

(c) Interpret the result.

Revision exercise 6

1 A small electrical shop recorded the yearly sales of radios (y) and television sets (x) over a period of 10 years. The results are shown in the table below.

Year	1	2	3	4	5	6	7	8	9	10
No. of televisions sold, x	60	68	73	80	85	88	90	96	105	110
No. of radios sold, y	80	60	72	65	60	55	52	44	42	36

(a) Using scales going from 50 to 120 for the sales of televisions and 30 to 90 for the sales of radios, draw a scatter diagram.

(b) What sort of correlation does the scatter diagram suggest?

(c) Is there a causal relationship between the television sales and radio sales?

2 In a woodland the number of breeding pairs of blackbirds (x) and the mean numbers of young blackbirds raised by each breeding pair (y) were recorded for 12 consecutive years. The table below shows the results.

Year	1	2	3	4	5	6	7	8	9	10	11	12
No. of breeding pairs, x	45	46	60	60	63	40	42	65	55	48	95	55
Mean number of young, y	6	7	5.5	4	1.8	4.5	6	5.2	4	4.5	1.3	2.5

It is suggested that there could be a negative correlation between the number of breeding pairs and the average number of young that were raised per breeding pair.

(a) Draw a scatter diagram for these data.

(b) How strong is the correlation?

(c) Draw by eye a line of best fit.

(d) Use your line of best fit to estimate the average number of young blackbirds per breeding pair if there were 50 breeding pairs.

3 It has been said that when the stock market in America goes up or down the stock market in London goes up and down with it. The table below gives the values of the stock market indices in the USA (Dow–Jones) and London (FTSE 100) at the end of eight consecutive weeks.

Week	1	2	3	4	5	6	7	8
USA index	8600	8400	8100	8200	7900	7750	8000	8010
London index	3800	3650	3500	3700	3600	3620	3720	3730

(a) Draw a scatter diagram for these data.

(b) Calculate the mean point and draw a line of best fit through it.

(c) Describe the correlation shown by the scatter diagram.

(d) Calculate the equation of the line of best fit.

(e) If the USA stock market was at 8500, what would you estimate the London stock market to be?

(f) If the USA stock market went up to 10 500 would you expect any estimate you make for the London stock market to be totally reliable? Give a reason for your answer.

4 The number of mites found in wheat grain was assessed every two days over a 12-day period. The results are shown in the table.

Day	0	2	4	6	8	10	12
Number of mites	100	180	300	550	900	1400	1900

(a) Draw a scatter diagram for these data.

(b) Suggest a suitable model for the relationship between these data.

5 The numbers of people (in thousands) engaged in the manufacture of motor vehicles and in the manufacture of accessories for motor vehicles over a period of nine years are shown in the table below.

Year	1	2	3	4	5	6	7	8	9
Manufacture of motor vehicles (1000s)	280	290	297	300	310	295	311	330	320
Manufacture of accessories (1000s)	90	95	110	125	140	145	155	170	175

(a) Rank these data.

(b) Calculate Spearman's correlation coefficient for these data.

(c) Comment on the strength of the correlation.

Summary of key points

1 You can use a scatter diagram to show whether two sets of data are related.

2 If the points on a scatter diagram lie approximately on a straight line, there is a linear relationship between the two sets of data.

3 Correlation is a measure of the strength of the linear association between two variables.

4 Negative correlation is when one variable tends to decrease as the other increases.

5 Positive correlation is when one variable tends to increase as the other one increases.

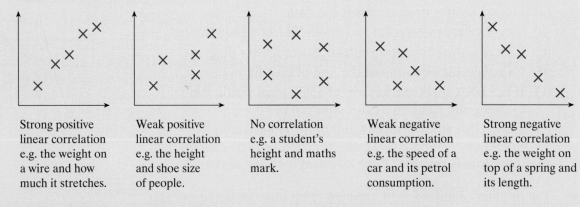

| Strong positive linear correlation e.g. the weight on a wire and how much it stretches. | Weak positive linear correlation e.g. the height and shoe size of people. | No correlation e.g. a student's height and maths mark. | Weak negative linear correlation e.g. the speed of a car and its petrol consumption. | Strong negative linear correlation e.g. the weight on top of a spring and its length. |

6 Association does not necessarily mean there is correlation.

7 When a change in one variable directly causes a change in another variable, there is a causal relationship between them.

8 Correlation does not necessarily mean there is a causal relationship.

9   A line of best fit is a straight line drawn so that the plotted points on a scatter diagram are evenly scattered either side of the line.

10 A line of best fit is a model for the association between two variables.

11 A line of best fit should pass through the mean point $(\bar{x}, \bar{y})$.

12 Interpolation is when you find values within the range of the values you are given.

13 Extrapolation is when you find values outside the range of the values you are given.

14 The equation of a line of best fit is $y = ax + b$, where a is the gradient of the line and b is the intercept with the y-axis.

$$a = \frac{y_2 - y_1}{x_2 - x_1} \text{ and } b = y_1 - ax_1 \text{ or } b = y_2 - ax_2.$$

15 Spearman's rank correlation coefficient (r_s) is given by

$$r_s = 1 - \frac{6\Sigma d^2}{n(n^2 - 1)} \text{ where } n = \text{number of sets of readings}$$

16 If r_s is close to 1 there is a strong positive linear correlation, and there is close agreement between the rankings.

17 If r_s is close to -1 there is a strong negative linear correlation, and there is close to complete disagreement between the rankings.

18 If r_s is close to zero then there is no linear correlation, and so no agreement or disagreement between the rankings.

7 Time series

7.1 Line graphs

In magazines and newspapers you will often
see graphs that look like this:

This is an example of a **line graph**.

- **A line graph is used to display data when
  the two variables are not related by an
  equation and you are not certain what
  happens between the plotted points.**

No reliable information can be found from the
graph for values lying between the points plotted.
A straight dotted line is used to join the points.

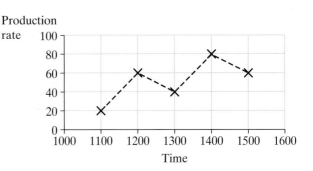

For example, on the graph above there
could be a tea break at 11.30 reducing
the production to zero for a short time.

Example 1

The table below shows the monthly rainfall at a seaside town
last year.

Month	Jan	Feb	Mar	Apr	May	Jun	Jul	Aug	Sept	Oct	Nov	Dec
Rainfall (cm)	18	21	12	16	11	9	6	10	7	15	18	23

Show these data on a line graph.

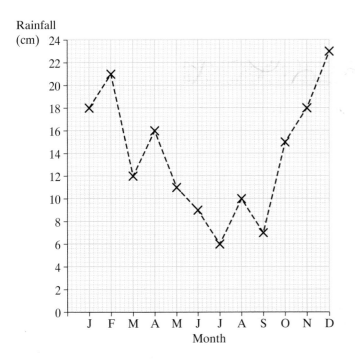

Each axis should have a scale that is
clearly marked and each axis should be
clearly labelled.

Because you do not know the rainfall
pattern between the points, joining them
has no real meaning, but it helps you to
see what is happening.

Plotting a graph

When plotting a graph, choose scales that fit your graph paper.
Your scales do not have to start at zero.

Example 2

This table shows the temperatures in degrees centigrade at different
times during part of a day in March. Draw a line graph of these data.

Time	0900	1000	1100	1200	1300	1400	1500
Temperature (°C)	10	11	13	14	15	12	11

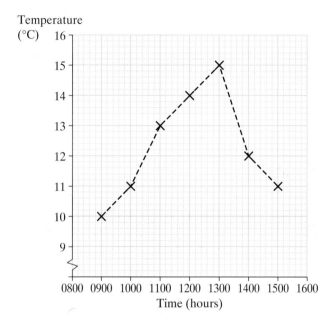

This graph has values only
between 9 and 15 on the
vertical axis. In this case the
scale starts at 8, which
makes it easier to plot and
read the values.

Exercise 7A

1 The graph shows the number of hours Joan watches television
   on each day of the week.

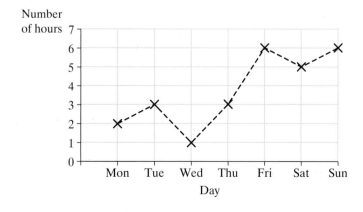

(a) How many hours does Joan watch television on Tuesdays?

(b) On which day does Joan watch least television?

(c) Suggest a reason why Joan watches television more on the last three days of the week.

2   The following table shows the amount of money Wing has in his bank account at the end of each month during the course of a year.

Month	Jan	Feb	Mar	Apr	May	Jun	Jul	Aug	Sept	Oct	Nov	Dec
Money (£)	50	65	78	84	100	96	84	24	38	43	78	48

(a) Draw a line graph of these data.

(b) There were two periods when Wing had large expenditures. In which months did these occur?

(c) At the end of which month did Wing have the greatest amount of money in his bank?

3   The table shows the number in thousands of people who died from flu over a period of ten years.

Year, x	1992	1993	1994	1995	1996	1997	1998	1999	2000	2001
Number of deaths in 1000s	3.8	1.5	4.6	3.4	16.0	3.2	6.8	3.4	4.2	3.0

(a) Draw a line graph of these data, starting your x-axis at 1991 and your y-axis at 0.

(b) Comment on the year 1996.

4   The table shows the wind speed in miles per hour during one day in May.

Time	0300	0600	0900	1200	1500	1800	2100	2400
Wind speed (m.p.h.)	10	15	20	25	20	25	30	35

(a) Draw a line graph of these data.

(b) Describe in words what happened to the wind speed between 03.00 and 24.00.

5   The table below shows the monthly profits made by a market stall selling vegetables.

Month	J	F	M	A	M	J	J	A	S	O	N	D
Profit (£100)	10	11	12	14	12	15	20	24	23	16	10	18

(a) Draw a line graph of these data.

(b) At what time of the year are profits at their highest?

7.2 Time series

In weather stations there are instruments that measure temperature and rainfall. Readings may be taken of these at a set time each day. These readings can be used to produce monthly rainfall figures.

Sets of observations of variables such as these taken over a period of time are called **time series**. In most cases the observations are made at equal time intervals.

■ **A time series is a set of observations of a variable taken over a period of time.**

A line graph can be used to show a time series. When plotting a time series, time is plotted on the horizontal axis.

Example 3

Shown below are the quarterly sales of ice cream over a period of time.

Year	1998		1999				2000				2001			
Quarter	3	4	1	2	3	4	1	2	3	4	1	2	3	4
Sales (£m)	3.6	1.4	2.0	3.6	4.6	2.6	3.0	4.6	5.6	3.4	3.6	5.2	6.3	4.6

(a) Draw a line graph of this time series.

(b) Comment on the graph.

(a)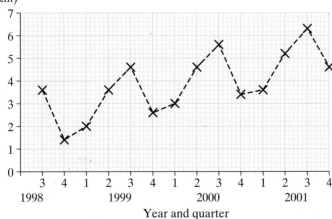

(b) The sales of ice cream are not the same throughout the year. The sales in quarter 1 of the year are quite low. Sales then increase in quarters 2 and 3 before dropping back in the fourth quarter of each year.

Exercise 7B

1 Which of the following could be shown on a time series?
   Explain your answer.
   (a) Sales of cold drinks.
   (b) Numbers of pets owned by a group of children.
   (c) Hours of sunshine.

2 Shown in the table are the quarterly rainfall figures for a town
   in the centre of Great Britain.

Year	2000		2001				2002				2003			
Quarter	3	4	1	2	3	4	1	2	3	4	1	2	3	4
Rainfall (cm)	8	21	26	14	8	20	24	13	4	19	19	11	3	20

   (a) Draw a line graph of this time series.
   (b) Comment on the graph.

3 The quarterly sales of computer play stations in a large store are
   shown below.

Year	2000	2001				2002				2003			
Quarter	4	1	2	3	4	1	2	3	4	1	2	3	4
Sales (100s)	14	2	3	3	12	3	4	5	10	4	5	6	9

   (a) Draw a line graph of this time series.
   (b) Comment on the graph.

4 The number of nurses at a large hospital who resigned from work
   was recorded. The result for a three-year period is shown below.

Year	2001				2002				2003			
Quarter	1	2	3	4	1	2	3	4	1	2	3	4
Number resigning	28	20	26	20	34	30	31	26	42	34	32	34

   (a) Draw a line graph of this time series.
   (b) In which quarter do most nurses resign?
   (c) Did nurses tend to resign more in 2003 compared to 2001?

5 A factory manager looked at the monthly profits on a particular
   manufacturing process over last year. The figures are shown in
   the table.

Month	J	F	M	A	M	J	J	A	S	O	N	D
Profit (£10 000)	14	16	15	15	17	16	15	18	19	18	19	20

   He thinks profits are falling.
   (a) Draw a line graph of this time series.
   (b) Comment on the manager's thoughts.

7.3 Trend lines

The graph below shows the sales of ice cream over three and a half years.

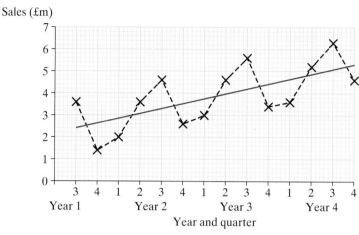

Although the sales seem to increase and decrease each year, the general *trend* is for the sales to rise. You can draw a **trend line** so that the plotted points are equally scattered either side of it.

■ **A trend line is a line that shows the general trend of the data.**

■ **A long-term trend is the way a graph appears to be going over a period of time. There may be a rising trend, a falling trend or a level trend.**

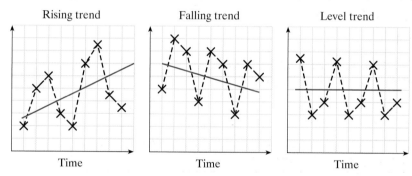

Exercise 7C

1 Which graph shows **(a)** a rising trend? **(b)** a level trend? **(c)** a falling trend?

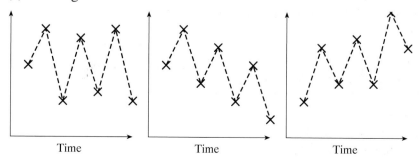

2 The table shows the quarterly sales of new cars by a garage over a three-year period.

Year	2001				2002				2003			
Quarter	1	2	3	4	1	2	3	4	1	2	3	4
Sales in 100s	10	3	8	5	14	7	13	9	16	10	18	9

(a) Copy this line graph and complete it by plotting and joining up the remaining points.

(b) Draw in a trend line.

(c) Comment on the trend of the data.

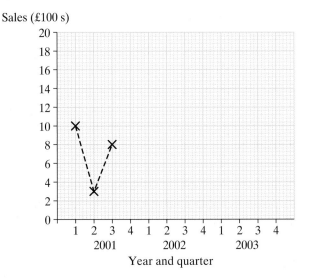

3 A company compiles quarterly sales figures. The figures for three years are shown below.

Year	Year 1				Year 2				Year 3			
Quarter	1	2	3	4	1	2	3	4	1	2	3	4
Sales in (£10 000)	45	48	50	52	40	25	22	45	28	30	25	36

(a) Draw a line graph of this time series.

(b) Draw in a trend line.

(c) Comment on the trend of these data.

4 A tour bus company organises trips to Scotland. They wish to find out if the tours are getting more or less popular so that they can plan for the future. Data for the past two years are shown in the table below.

Year	Year 1						Year 2					
Months	Jan–Feb	Mar–Apr	May–Jun	Jul–Aug	Sep–Oct	Nov–Dec	Jan–Feb	Mar–Apr	May–Jun	Jul–Aug	Sep–Oct	Nov–Dec
Number of people on trip	24	38	40	52	40	22	20	36	30	46	30	20

(a) Draw a line graph of this time series.

(b) Draw in a trend line.

(c) Comment on the trend of the data.

7.4 Variations in a time series

The variations in a time series may be due to two things.

1. **A general trend**: The trend line shows the trend of a time series. This may be increasing, decreasing or staying level.

2. **Seasonal variation**: Many industries do not experience an even demand for their products.

> Many shops have their greatest sales just before Christmas: their sales show a yearly seasonal variation that peaks at Christmas.

The graph on page 187 shows the sales of ice cream for a particular company over three and a half years.

The trend of ice cream sales is upwards. The sales in each of the first quarters of the year are less than the trend would suggest.
Those in each of the second quarters are above the trend. Those in each of the third quarters are well above the trend.
The fourth quarter sales are below the trend.
This is because ice cream sells better in the warmer months and less well in the colder months.

■ **Seasonal variations are the differences between actual and trend values shown by the trend line. They have a pattern that repeats each year.**

Example 4

The table shows quarterly figures over three years for the number of unemployed building workers. The data are in thousands.

Year	1999				2000				2001			
Quarter	1	2	3	4	1	2	3	4	1	2	3	4
Unemployed builders (1000s)	20	13	13	17	19	12	12	14	18	9	9	12

(a) Draw a time series graph of these data.
(b) Draw a trend line on the graph.
(c) Describe the variations that are taking place.

(a) (b)

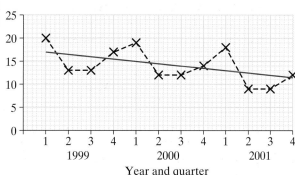

(c) The graph shows a falling trend – the number of builders out of work tends to fall each year. There is a seasonal variation with the unemployment above the trend value in the first quarters and lower than the trend values in the second and third quarters.

Exercise 7D

1 Which of the following is likely to show seasonal variations?
Explain your answer.
   (a) The number of bank accounts opened.
   (b) The number of hours of sunshine.
   (c) The sales of swimsuits.
   (d) The sales of breakfast cereals.

2 The graph below shows the quarterly sales of hot-dogs from a
market stall during three consecutive years.

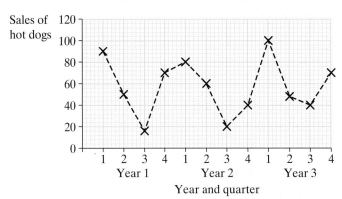

   (a) Copy this graph and draw a trend line.
   (b) The sales seem to go up and down about the trend line.
        Give a reason for this.
   (c) In which quarter of the year are the sales highest?

3 The line graph below shows the number of people going on a
particular coach tour in each quarter of the years 1999 to 2002.

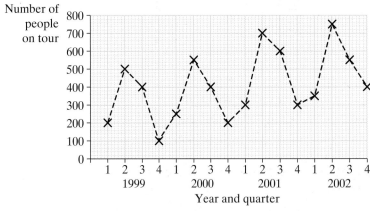

   (a) How many people went on the tour during the second
        quarter of 1999?
   (b) How many people went on the tour during the whole of 1999?
   (c) What was the general trend of the data?
   (d) Comment on the seasonal variations.

4 The table shows the quarterly sales of hot baked potatoes at a stall in a busy market square over 3 years.

Year	2002				2003				2004			
Quarter	1	2	3	4	1	2	3	4	1	2	3	4
Sales (£1000)	14	10	2	8	9	7	4	12	11	6	3	11

(a) Draw a line graph for these data.

(b) Draw in a trend line.

(c) Comment on the trend and seasonal variations.

7.5 Using ICT to draw line graphs

Example 5

The consumption of beer (in millions of gallons) over an 8-year period is shown.

Year, x	1	2	3	4	5	6	7	8
Consumption (millions of gallons)	15	10	22	25	15	20	30	23

(a) Draw a line diagram.

(b) Add a trend line.

(a) Using the Microsoft Excel worksheet button on the tool bar.

	A	B	C
1	Year	Consumption	
2	1	15	
3	2	10	
4	3	22	
5	4	25	
6	5	15	
7	6	20	
8	7	30	
9	8	23	

1. Enter information into new spreadsheet.
   - In cell A1 type 'Year', in B1 type 'Consumption'.
   - In cell A2 type '1', in A3 type '2', in A4 type '3'.
   - Highlight cells A2, A3 and A4.
   - Place the curser at the bottom right hand corner of cell A4. Hold down the left hand button on the mouse and drag down the column to enter the years up to 8. This copies the year pattern into the remaining cells.
   - In cell B2 type '15', in B3 type '10', and so on until all the consumptions have been entered.

2. Highlight all the cells you have used.

3. Use the **chart wizard** button on the tool bar.
   - (Step 1 of 4) From chart type select **XY(scatter)**.
   - From chart sub type select the bottom left diagram.

- (Step 2 of 4) Click **Next** button.
- (Step 3 of 4) Chart options.
  In Value (X) axis box, enter 'Year'.
  In Value (Y) axis box, enter
  'Consumption (million gallons)'.
  Click **Next** button.
- (Step 4 of 4) Click **Finish** and the graph is obtained.

The chart looks like this:

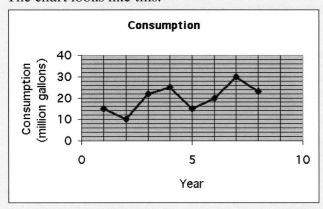

4. To change scales.
   - Click on the *x*-axis.
   - Click **Format** on toolbar, then **Select axis** then **Scale**.
   - Change *x*-axis (year) to Major unit 1. Click **OK**.
   - Click on *y*-axis (consumption).
   - Click **Format** on toolbar, then **Select axis**.
   - Change *y*-axis to Major unit 10. Click **OK**.
   - Click on the points, then **Format** on the toolbar.
     Change line style to dotted.

   The graph that results is as follows.

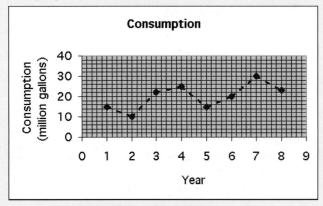

(b) To add a trend line to your chart:
   - Double click on the chart.
   - Click **Chart** on the toolbar.
   - On the drop down menu click on **Add trend line**.

Exercise 7E

1   The table shows the monthly sales, in £1000s, of a shop over a one-year period.

Quarter	1	2	3	4	5	6	7	8	9	10	11	12
Sales (£1000s)	33	38	34	39	35	39	35	39	37	41	39	43

Use ICT to draw a line graph and trend line for these data.

7.6 Moving averages

Example 6

The graph shows Nassim's scores on a computer game over a period of time.

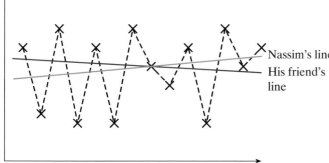

Nassim has drawn a trend line on the graph to show that his score is improving, but his friend has drawn a second one showing that he is getting worse. Who is right? Is Nassim improving or is he getting worse?

In this instance it looks as if Nassim's line is nearer to the truth.

As in this example, the line chart of a time series may show large variations with upward and downward peaks. This can make it difficult to see the trend of the data and thus to draw a trend line.

A good way of seeing the trend is to use **moving averages**. Moving averages are the averages of successive observations of a time series. The averaging may be done over 2, 3, 4, 5, 6, etc., readings and the averages are known as the 2 point, 3 point, 4 point, 5 point, 6 point, etc. moving averages.

There will be a maximum of 7 points in an exam.

   A moving average is an average worked out for a given number of successive observations as you work through the data.
   A 3 point moving average uses three observations.
   A 4 point moving average uses four observations, and so on.

Example 7

Quarterly car sales over a two-year period are shown in the table.

Quarter	1	2	3	4	5	6	7	8
Cars sold/quarter	16	26	30	24	28	36	35	42

Find the 4 point moving averages, and plot them on a time series.

The first moving average would be $\dfrac{16 + 26 + 30 + 24}{4} = 24$.

The second moving average would be $\dfrac{26 + 30 + 24 + 28}{4} = 27$.

The third moving average would be $\dfrac{30 + 24 + 28 + 36}{4} = 29.5$.

Proceeding in the same way the next two moving averages are 30.75 and 35.25.

Moving averages are plotted at the mid-point of the time intervals they cover. Do not join up the points plotted for the moving averages.

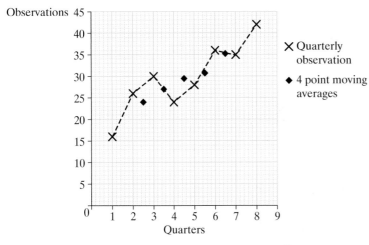

X Quarterly observation

◆ 4 point moving averages

The eight points produce five moving averages.

The first will be the mid-point of quarters, 1, 2, 3 and 4 = 2.5.
The second would be at the mid-point of 2, 3, 4 and 5 = 3.5.

The moving averages clearly show an upward trend.

Moving averages even out seasonal variations, and this makes any trend easier to see. The number of points in each moving average should cover one complete cycle of seasons (this ensures that each moving average contains one reading from each season).

Example 8

The following table gives the takings (in £1000s) of a shopkeeper in each quarter of three successive years.

Year	Quarter			
	1st	2nd	3rd	4th
1	15	25	53	24
2	15	27	59	26
3	17	28	60	25

(a) Draw a line graph to illustrate these data and on the same graph show the 4 point moving averages.

(b) Give the reason why a 4 point moving average is used in this case.

(c) State the conclusion that may be drawn from the graph.

(a) Calculating the moving averages is best done by making a table.

Year	Quarter	Takings	4 point moving average
1	1	15	
1	2	25	(15 + 25 + 53 + 24)/4 = 29.25
1	3	53	(25 + 53 + 24 + 15)/4 = 29.25
1	4	24	(53 + 24 + 15 + 27)/4 = 29.75
2	1	15	(24 + 15 + 27 + 59)/4 = 31.25
2	2	27	(15 + 27 + 59 + 26)/4 = 31.75
2	3	59	(27 + 59 + 26 + 17)/4 = 32.25
2	4	26	(59 + 26 + 17 + 28)/4 = 32.5
3	1	17	(26 + 17 + 28 + 60)/4 = 32.75
3	2	28	(17 + 28 + 60 + 25)/4 = 32.50
3	3	60	
3	4	25	

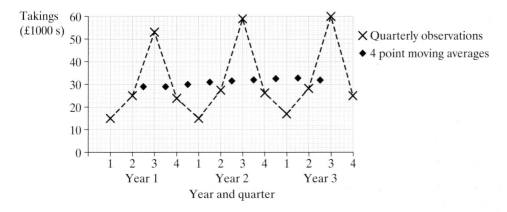

(b) A 4 point moving average is used because the seasonal changes take place over four quarters.

(c) The general trend is upwards. A seasonal variation is present. The takings are above trend in the third quarter of the year and below trend in the first quarter.

Drawing a trend line through moving averages

You can draw a trend line through the moving averages by eye. This will be more accurate as a measure of trend than one drawn using the original data.

The diagram on the next page shows the line graph for Example 8 with a trend line added for the moving averages.

You will not be asked to draw your trend line through a mean point.

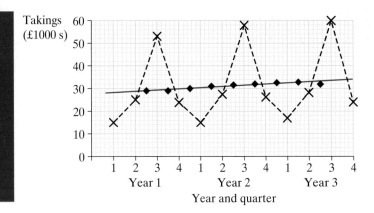

Takings (£1000 s) — Year and quarter

Exercise 7F

1 The following table gives the monthly values of a company's production, in thousands of items, from January 1999 to February 2000. Calculate the 3 point moving averages.

Month	Jan	Feb	Mar	Apr	May	Jun	Jul
Production	108	83	100	108	111	115	106

Month	Aug	Sep	Oct	Nov	Dec	Jan	Feb
Production	103	118	119	124	114	120	125

2 **(a)** Explain what is meant by a 4 point moving average.

The following table gives the numbers (in millions of pounds) of exports from Great Britain to a continental country between July 2000 and June 2001.

Month	Jul	Aug	Sep	Oct	Nov	Dec
Exports	9.8	10.2	9.7	11.4	10.5	10.7

Month	Jan	Feb	Mar	Apr	May	Jun
Exports	11.4	10.9	12.6	11.3	13.3	11.8

(b) Draw a line graph of these figures.

(c) Calculate the 3 point moving averages and add these to your graph.

3 The following table shows the number of new houses completed in a city in 12 consecutive quarters.

	First quarter	Second quarter	Third quarter	Fourth quarter
First year	260	285	200	265
Second year	240	250	270	280
Third year	280	270	290	295

(a) Draw a line graph of these data.

(b) Calculate and plot the appropriate moving averages.

(c) Is the trend rising, falling or level?

4   The following table gives the annual circulation of library books (in thousands) for the library in a small town.

Year	1990	1991	1992	1993	1994	1995
Circulation (thousands)	14.5	16.5	14.0	14.5	18.0	15.5

Year	1996	1997	1998	1999	2000	2001
Circulation (thousands)	14.5	15.0	17.0	16.0	19.5	18.5

(a) Draw a line graph to illustrate these figures and add the three-yearly moving averages.

(b) Comment on the trend.

5   The half yearly profits, in £1000s, made by a computer shop are shown in the table.

Year	1	2	3	4	5
Jan–Jun	18	18	22	24	26
Jul–Dec	22	26	26	28	34

(a) Plot these data on a time series graph.

(b) Calculate appropriate moving averages and plot them on the graph.

(c) Draw a trend line by using the moving averages. Comment on the trend.

6   The average weekly sales (in £1000) for two competing firms over a three-year period are shown in the table.

Year		1998			1999			2000		
Months		Jan–Apr	May–Aug	Sep–Dec	Jan–Apr	May–Aug	Sep–Dec	Jan–Apr	May–Aug	Sep–Dec
Firm	A	280	260	310	480	450	530	730	710	740
	B	480	360	420	500	430	480	640	520	550

(a) Plot both sets of data on the same graph.

(**b**) Calculate appropriate moving averages for both firms to eliminate the seasonal variation in sales.

(**c**) Plot the moving averages on your graph.

(**d**) Draw trend lines for the moving averages for both firms.

(**e**) Use your trend lines to estimate in which period of time the sales of firm A first equalled the sales of firm B.

7.7 Using ICT to calculate moving averages

Example 9

Calculate the moving averages for Example 8 using a spreadsheet.

1. Enter information into new spreadsheet.

- In cell A1 type 'Quarter', in B1 type 'Moving average' and in C1 type 'Takings'.
- In cell A2 type '1', in A4 type '2', in A6 type '3'. Highlight cells A2 to A7.
- Place the cursor at the bottom right hand corner of cell A4. Hold down left hand button on the mouse and drag down the column to enter the quarters up to cell A25. This copies the quarters pattern into the remaining cells.
- In cell B2 type '15', in B4 type '25', and so on until all the Takings have been entered.

	A	B	C
1	Quarter	Moving average	Takings
2	1	15	
3			
4	2	25	
5			29.25
6	3	53	
7			29.25
8	4	24	
9			29.75
10	5	15	
11			31.25
12	6	27	
13			31.75
14	7	59	
15			32.25
16	8	26	
17			32.5
18	9	17	
19			32.75
20	10	28	
21			32.5
22	11	60	
23			
24	12	25	

2. Calculate the moving averages.

- In C5 enter the formula '=(B2+B4+B6+B8)/4' and click on the tick button. (Do not leave spaces between letters, numbers and signs.)
- Place the cursor at the bottom right corner of cell C6. Hold down the left hand button on the mouse and drag down the column to enter down to cell C21.
- The moving averages will appear in the appropriate boxes.

Note: with a 3 or 5 point moving average the process is simpler, as the moving average does not go between the points.

Exercise 7G

1   The following table gives the monthly totals (in 100s) of new
    motorcycles bought during 2002.

Month	Jan	Feb	Mar	Apr	May	Jun
Motorcycles	120	90	105	120	85	85

Month	Jul	Aug	Sep	Oct	Nov	Dec
Motorcycles	90	45	50	55	20	25

(a) Draw a line graph to illustrate these data.

(b) Use a spreadsheet to calculate the 3 point moving
    averages and add them to the graph.

(c) Comment on the trend.

7.8 Estimating seasonal variations

The seasonal variation for a point on the graph is given by:

■   **Seasonal variation at a point = actual value − trend value**

Mean seasonal variations

Seasonal variations each year for a given quarter vary from year to
year. An estimate of a particular quarter's seasonal variation is
found by:

■   **Estimated mean seasonal variation = mean of all the seasonal
    variations for that season.**

Example 10

The quarterly sales of a company over a three-year period (in
millions of pounds) are shown below. Plot the moving averages on a
time-series graph, and calculate the mean seasonal variations.

Year	Year 1				Year 2				Year 3			
Quarter	1	2	3	4	1	2	3	4	1	2	3	4
Sales (£1 000 000s)	14	9	16	21	18	11	21	23	18	15	25	27

For quarterly seasonal variations a 4 point moving average is used to
find the best trend line.

Year and quarter		Actual sales (£m)	4 point moving average
1	1	14	
1	2	9	
			15.00
1	3	16	
			16.00
1	4	21	
			16.50
2	1	18	
			17.75
2	2	11	
			18.25
2	3	21	
			18.25
2	4	23	
			19.25
3	1	18	
			20.25
3	2	15	
			21.25
3	3	25	
3	4	27	

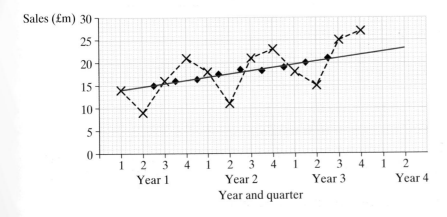

The seasonal variations at each plotted point are found by subtracting the readings at the trend line from the quarterly sales at the same point.

Year and quarter		Actual sales (£m)	Trend (from trend line)	Seasonal variation at a point = actual − trend
1	1	14	14.0	0
	2	9	14.7	−5.7
	3	16	15.5	+0.5
	4	21	16.2	+4.8
2	1	18	16.9	+1.1
	2	11	17.6	−6.6
	3	21	18.4	+2.6
	4	23	19.1	+3.9
3	1	18	19.8	−1.8
	2	15	20.6	−5.6
	3	25	21.3	+3.7
	4	27	22.0	+5.0

To calculate the estimated mean seasonal variation:

Year	Quarter			
	1	**2**	**3**	**4**
1	0	−5.7	+0.5	+4.8
2	+1.1	−6.6	+2.6	+3.9
3	−1.8	−5.6	+3.7	+5.0
Totals	−0.7	−17.9	+6.8	+13.7
Estimated mean seasonal variation	$\frac{-0.7}{3} = -0.23$	$\frac{-17.9}{3} = -5.97$	$\frac{6.8}{3} = +2.27$	$\frac{13.7}{3} = +4.57$

Looking at the table you can see that the estimated mean seasonal variation for the third quarter is +2.27 million pounds while that for the second quarter is −5.97 million pounds.

7.9 Making predictions

A trend line and the estimated mean seasonal variations can be used to predict the sales figures at some future time.

You often have to extend the trend line beyond the plotted values.

■ **Predicted value = trend line value (as read from trend line on graph) + estimated mean seasonal variation**

For example, in Example 10 the predicted actual sales figures for the first quarter of year 4 would be:

Predicted actual sales = Trend line value (as read from trend line on
the graph) + estimated mean seasonal
variation
= 22.7 − 0.23
= 22.47 million pounds

For the second quarter
predicted actual sales = 23.5 − 5.97
= 17.53 million pounds

Accuracy of predictions

The accuracy of any prediction will depend upon two things:

● how far into the future the prediction is made – the further into
the future the prediction is made the less accurate it will be;

● how good the estimates of the mean seasonal variations are.

Exercise 7H

1   Given a trend line value of £13.39 and a mean seasonal
variation of £3.20 work out the predicted value.

2   The following table shows the seasonal variations in the price of
lettuce. Copy and complete the diagram.

Year	Seasonal variation			
	1	2	3	4
1	0.0	−6.2	8.1	4.0
2	3.1	−4.0	5.0	3.5
Total				
Mean seasonal variation				

3   The following table shows actual and trend values of the sales of
mountain bicycles at a small shop. Copy and complete the table.

Year and quarter		Actual value	Trend	Seasonal variation
1	1	26	22	
1	2	38	24	
1	3	14	18	
1	4	10	14	
2	1	20	19	
2	2	32	23	
2	3	13	17	
2	4	10	10	

4   Using the values in question 3, draw up a table and work out the estimated mean seasonal variation.

5   The table shows the quarterly profits of a factory (in £1000s) for the years 2000 to 2002.

Year	Quarter			
	1	2	3	4
2000	30	54	60	40
2001	46	90	92	60
2002	74	122	132	96

The information for the years 2000 and 2001 has been plotted as a line graph.
The first five 4 point moving averages have been plotted on this graph. They are 46, 50, 59, 67, 72.

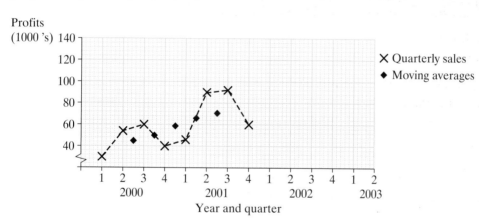

(a)   Copy this graph and complete it by plotting the remaining points and the moving averages.

(b)   Draw the trend line and extend it to the first quarter of 2003.

(c)   Given that the mean seasonal trend for the first quarter is −14 estimate the profit for the first quarter of 2003.

6   The table shows the quarterly electricity costs (in £s) for a house over a two-year period.

(a)   Plot these data on a line graph.

(b)   Work out the 4 point moving averages and plot these on the graph.

Year	Quarter			
	1	2	3	4
1	180	142	60	110
2	196	146	76	122

(c)   Draw in a trend line using the moving averages.

(d)   Work out the mean seasonal variations for the first two quarters of the year.

(e)   Predict the electricity bills for the first two quarters of year 3.

7 (a) Describe what is meant by seasonal change and give an example of this.
       The table shows company A's quarterly sales (in £1000s).

	Quarter			
Year	**1**	**2**	**3**	**4**
1998			120	100
1999	86	138	132	92
2000	90	122	128	84
2001	110	122	140	80
2002	118	130		

   (b) Plot the sales and moving averages on a graph, and add a trend line.
   (c) Work out the mean seasonal variations.
   (d) Predict the sales for the third and fourth quarters of 2002.

8   The table below shows the number of articles sold over a three-year period by a manufacturing company.

	Quarter			
Year	**1**	**2**	**3**	**4**
1999	686	590	660	720
2000	754	642	732	808
2001	842	738	808	900

   (a) Work out the 4 point moving averages.
   (b) Using a scale of 1 cm to represent 20 articles on the vertical axis, and starting at 500, represent the data and moving averages on a line graph.
   (c) Draw by eye a line to represent the trend of the graph.
   (d) By taking readings from your graph draw up a table to work out the mean seasonal variations.
   (e) Use your result in (d) to estimate the first two quarterly figures for 2002.

7.10 Calculating the equation of the trend line

A law of the form $y = ax + b$ can be fitted to the trend line using the same method as you have used in the last chapter.

■ **The equation of a trend line is $y = ax + b$, where a is the gradient of the line and b is the intercept with the y-axis.**

Before you can work out the equation, the periods on the x-axis must be labelled 1, 2, 3, etc. (Measures such as '2002 quarter 1' are difficult to deal with.)

The constant a is the amount by which the trend increases per time period (quarter, two-months, etc). The constant b is the value of the trend at the end of the last period before observation began.

Remember: $a = \dfrac{y_2 - y_1}{x_2 - x_1}$

and $b = y_1 - ax_1$ or $b = y_2 - ax_2$.

When calculating predicted values, you can use the equation of the trend line to work out a particular trend line value instead of reading it off the graph.

■ **Predicted value = trend line value (as read from trend line on graph, or calculated by law) + seasonal variation**

Example 11

A factory keeps records of the number of days work lost through illness. The records for two years are summarised in the table. Each figure represents the number of days lost over a two-month period.

Year	Number of days lost					
1999	222	246	168	130	100	184
2000	255	273	198	154	133	187

(a) Draw a graph to illustrate these figures and on it superimpose moving averages using an appropriate number of observations.

(b) Draw a trend line on the graph. Find the equation of this line, and give an interpretation of it.

(a) Use a 6 point moving average and number the two-month periods from 1 to 12.

Two-month period	Days lost	Moving average
1	222	
2	246	
3	168	
		175
4	130	
		180.5
5	100	
		185
6	184	
		190

Two-month period	Days lost	Moving average
7	255	
		194
8	273	
		199.5
9	198	
		200
10	154	
11	133	
12	187	

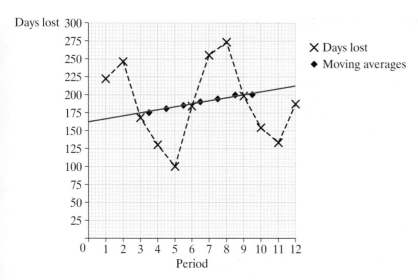

(b) If y = days lost and x = period, the law will be $y = ax + b$.

From the graph two points on the trend line are $(x_1, y_1) = (3, 175)$ and $(x_2, y_2) = (9, 200)$.

So $a = \dfrac{y_2 - y_1}{x_2 - x_1} = \dfrac{200 - 175}{9 - 3} = \dfrac{25}{6}$

and $b = y_2 - \dfrac{25}{6}x_2 = 200 - \dfrac{25}{6} \times 9 = 162.5$

The equation is $y = \frac{25}{6}x + 162.5$

The trend is for the number of days lost by illness to increase by $\frac{25}{6}$ over each two-month period. The estimated trend value was 162.5 at the last period before recording started.

> Alternatively, b can be read off the y-axis at period = 0. In this case this would not be very accurate because the scale is small.

Exercise 7I

1 The diagram shows a trend line.

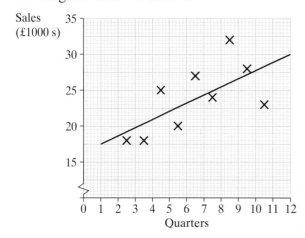

(a) Work out the gradient of the line.

(b) Describe in words what this gradient represents.

2   The table shows the turnover (in £1000s) of a department store.

Year	Four-month period		
	Jan–Apr	**May–Aug**	**Sep–Dec**
1999		134	162
2000	142	149	180
2001	166	182	210
2002	196		

(a) Plot the sales on a line graph.

(b) Work out appropriate moving averages and add these to your graph. Draw a trend line using the moving averages.

(c) Work out the equation of your trend line, using May–Aug 1999 as $x = 1$.

(d) Work out the mean seasonal variations.

(e) Predict the sales for the second and third four-month periods of 2002.

3   The number of pairs of size 6 trainers sold by a shoe shop is shown.

Year	Period	Trainers sold
1	Jan–Apr	49
	May–Aug	130
	Sep–Dec	70
2	Jan–Apr	40
	May–Aug	121
	Sep–Dec	55
3	Jan–Apr	31
	May–Aug	112
	Sep–Dec	31

(a) Plot these data on a time series graph.

(b) Explain why a 3 point moving average should be used for these data.

(c) Work out the 3 point moving averages and plot them on your graph.

(d) Draw a trend line.

(e) Work out the gradient of the trend line and give an interpretation of this.

(f) Work out the mean seasonal variations.

(g) Predict the number of size 6 trainers that will be sold in the three periods of year 4.

7.11 Quality assurance

A packet of crisps must have the weight of the contents marked on it. It might, for example, be marked 50 g. The manufacturers try to keep the weight as near as possible to this **target value**.

When products are made on a production line basis there will be some variation in the size or weight (**quality**) of the product.

There is bound to be some difference in the weights of crisps in a packet as a result of the manufacturing process. These changes are unavoidable. The weight of crisps in any packet may not be exactly 50 g; however, the mean weight of the packets produced should stay at 50 g and the difference between the minimum and maximum weights should stay constant.

Changes in the size/weight can also be caused by wear of tools, etc. These changes will cause either the mean weight to change, or the range of the weights to change. If a change of this type occurs the process is stopped. Quality assurance gives warning of such changes.

To keep check on the quality of the product, samples are taken at regular time intervals, and time series charts constructed for both the sample mean and the sample range.

■ **A control chart is a time-series chart that is used for process control.**

Quality control charts for means

The diagram shows a control chart used for the mean weight of a packet of crisps.

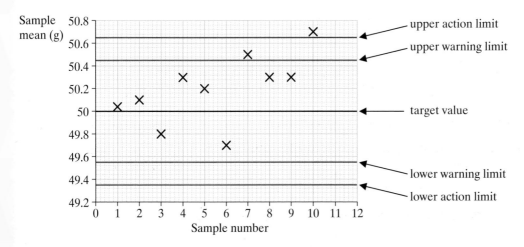

A line on the chart indicates the target value of 50 g for the mean weight.

Warning limits are set so that 95% of the samples should lie between them.

Action limits are set so that all the means will lie within them.

If a sample mean is between the warning limits the process is in control and the product is acceptable.

If a sample mean lies between the warning and action limits another sample is taken immediately.

If a sample mean is outside the action limits the process is stopped and the machinery is reset.

In practice this will be 99.8%.

In the chart above, sample 7 would cause another sample to be taken immediately. Sample 10 would cause the process to stop.

Quality control chart for medians

In some circumstances the median is plotted instead of the mean. The median is easier to find, and can be plotted more quickly. The control chart for medians looks the same as the control chart for means.

Quality control chart for ranges

The range of a product depends on the machinery being used and cannot be set. There is no target value for the range.

Quality control charts for ranges have action and warning limits the same as the charts for means.

Ideally the range would be zero, but you would be suspicious if the range was too low.

A control chart for sample ranges of the weight of crisp packets is shown below.

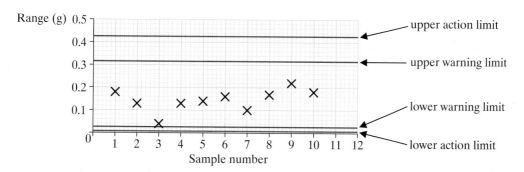

Sample ranges within the warning limits are acceptable.

Sample ranges between the warning and action limits would cause another sample to be taken.

If any range is outside the warning limits the process is stopped.

Sometimes the lower warning and action limits on a range chart are omitted.

The range in this instance is under control, since all points lie within the warning limits.

■ **If the plotted values are within the warning limits the process is under control. If a value is between the warning and action limits another sample is taken. If it is outside the action limits the process is stopped and the machine is reset.**

Example 12

The following samples were taken from a machine producing pins whose target length was 5.04 cm. The mean of the samples has to be within the range 4.98 to 5.1 cm, with warning limits at 5.08 and 5.0 cm, and the range has to be less than 0.3 cm with a warning limit at 0.25 cm. The process was reset after sample 4.

Sample	1	2	3	4	5	6	7	8	9	10
Size	4.94	5.17	5.02	5.16	5.03	5.09	4.93	5.11	4.97	5.00
	5.06	5.01	5.03	5.03	5.13	5.10	5.15	5.05	5.19	5.16
	5.12	5.03	4.98	5.14	4.99	4.99	5.10	4.90	5.05	5.02

(a) Work out the means and ranges of the samples and plot them on control charts.

(b) Comment on the charts.

(a) The means and ranges of the samples are:

Sample	1	2	3	4	5	6	7	8	9	10
Mean	5.04	5.07	5.01	5.11	5.05	5.06	5.06	5.02	5.07	5.06
Range	0.18	0.16	0.05	0.13	0.14	0.11	0.22	0.21	0.22	0.16

Control chart for means

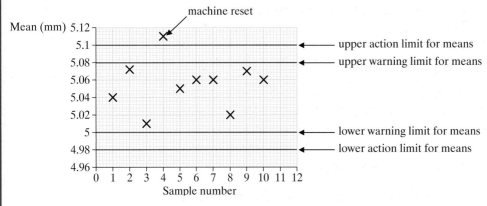

Control chart for range

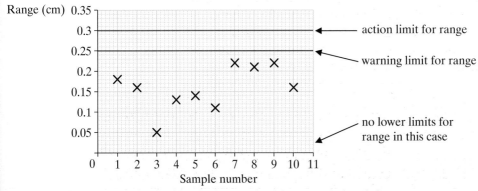

The sample mean was too large at sample 4 but came back into control after the machine was reset. The range was under control.

Exercise 7J

1 What is a control chart used for?

2 Explain why it is necessary to have control charts for both mean and range.

3 A machine is used for cutting curtain rails to a length of 150 cm. Samples of size 4 are taken at regular time intervals. The means of the samples have to be within the limits 149.2 cm and 150.8 cm, and their ranges have to be less than 3.28 cm. The first eight samples are shown.

Sample	1	2	3	4	5	6	7	8
Length (cm)	150.3	150.2	150.1	149.2	149.8	149.3	150.2	149.3
	149.9	149.7	149.7	149.7	150.7	149.7	150.0	150.7
	150.7	149.2	149.2	150.6	150.2	150.1	149.7	150.6
	150.7	150.9	150.2	150.9	149.3	149.9	149.1	149.4

Plot control charts for the mean and range.

4 A machine is being used to fill packets with crisps. The target weight of the packets is 50 g. Samples of size 3 are taken at regular time intervals. The sample means have to be within the limits 49 g to 51 g. The range has to be less than 2 g. The first eight samples are shown.

Sample	1	2	3	4	5	6	7	8
Weight (g)	49.02	50.10	49.05	50.02	50.04	50.00	49.90	50.02
	50.02	50.03	49.88	50.00	50.00	49.88	49.70	49.89
	49.70	49.81	49.90	49.77	49.96	48.38	50.10	49.40

Plot control charts for the mean and range.

5 A manufacturer of electrical shafts wishes to control the diameter of the shafts. At the end of each hour a sample of four shafts is taken and the diameter of the shafts measured. The target value of the shaft diameters is 38 mm. The action limits for the means of the samples are 37.6 and 38.4 cm, the warning limits are 38.27 and 37.73 cm. The ranges of the samples have to be between 0.04 and 1.247 cm. The first ten samples are given in the table.

Sample	1	2	3	4	5	6	7	8	9	10
Diameter (cm)	38.27	38.40	36.88	38.24	37.33	37.52	38.82	38.90	39.22	39.50
	37.98	38.22	37.88	38.05	38.00	38.18	37.98	38.66	38.88	39.42
	37.70	37.66	37.84	38.01	38.56	38.33	37.88	38.32	39.91	39.88
	37.17	37.88	38.00	37.86	37.91	37.89	38.36	38.88	38.99	40.36

Plot control charts for mean and range. Comment on your charts.

Revision exercise 7

1   The total monthly turnovers of the engineering industry in England over six consecutive months are shown in the table.

Month	Jul	Aug	Sep	Oct	Nov	Dec
Turnover (£ billion)	6.4	6.0	7.0	6.5	6.7	6.4

Draw a line graph of these data.

2   The quarterly takings of a post office in £1000s are shown in the table for three successive years.

Year	Quarter 1	2	3	4
1	26	42	46	74
2	26	40	47	76
3	34	48	56	90

(a) Draw a line graph of this time series.
(b) Draw a trend line on your graph.
(c) Comment on the general trend as shown by your trend line.
(d) Does the graph suggest that there is a seasonal variation? If so in which quarter are the sales highest?
(e) Give a reason why sales might be higher in the last quarter of the year.

3   A multiplex cinema shows a set of films for three consecutive weeks instead of for one week only. The attendances (in 100s) are shown below.

Day	Week 1	2	3
Mon	6	4	4
Tues	6	5	5
Wed	8	7	6
Thur	11	9	6
Fri	14	12	12
Sat	25	20	18
Sun	20	16	15

(a) Draw a line graph for this time series.
(b) Calculate 7 point moving averages and add them to your graph.
(c) Comment on the trend and seasonal variations.

4 The table shows the numbers of new houses (in 100s) completed in southern England over 12 consecutive quarters.

Year	Quarter 1	2	3	4
1	186	182	220	240
2	170	180	250	280
3	190	200	260	300

(a) Draw a line graph of this time series.

(b) Calculate the 4 point moving averages and plot them on the graph.

(c) Draw a trend line through the moving averages.

(d) Calculate the equation of the trend line.

(e) Calculate the mean seasonal variations.

(f) Using your answers to (d) and (e) predict the number of houses that will be completed in the second quarter of year 4.

5 The assistant manager of a clothes factory is anxious to ensure that the sizes of skirt waistbands in a large contract are made to the correct size. In order to do this he proposes to use control charts for mean and range.

(a) If one of the plotted points falls between a warning and an action limit what action should the manager take?

(b) If one of the plotted points falls outside the action limits what action should the manager take?

The target size of the waistbands is 26 cm measured from one side of the skirt waistband to the other when the skirt is laid flat. The manager decides to take samples of four skirts at a time. The first eight samples are shown in the table.

Sample	1	2	3	4	5	6	7	8
Size (cm)	25.6	26.8	25.0	25.6	24.9	26.2	25.9	25.2
	25.8	25.3	26.0	26.2	26.3	25.7	25.5	25.8
	26.4	26.3	25.4	25.3	26.2	24.7	26.5	25.3
	25.4	26.0	26.4	26.1	26.2	26.2	25.3	26.5

(c) Using warning limits of 26.39 and 25.61 cm, and action limits of 26.62 and 25.38 cm, draw a control chart for the mean and plot the sample means on it.

(d) Using warning limits of 0.24 and 1.59 cm and action limits of 0.08 and 2.12 cm draw a range control chart and plot the sample range on it.

(e) Comment on the state of the process.

Summary of key points

1   A line graph is used to display data when the two variables are not related by an equation and you are not certain what happens between the plotted points.

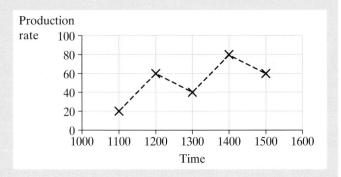

2   A time series is a set of observations of a variable taken over a period of time.

3   A trend line is a line that shows the general trend of the data. It can be drawn by eye.

4   A long-term trend is the way a graph appears to be going over a long period of time. There may be a rising trend, a falling trend or a level trend.

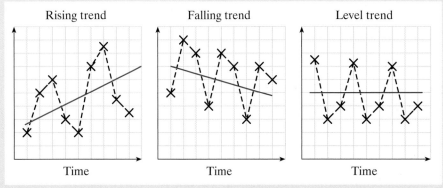

5   Seasonal variations are the differences between actual and trend values shown by the trend line. They have a pattern that repeats each year.

6   A moving average is an average worked out for a given number of successive observations as you work through the data.
A 3 point moving average uses three items of data.
A 4 point moving average uses four items of data and so on.

7   Seasonal variation at a point = actual value − trend value.

8   Estimated mean seasonal variation = mean of all the seasonal variations for that season.

9   Predicted value = trend line value (as read from trend line on graph, or calculated by the law) + estimated mean seasonal variation.

10 The equation of a trend line is $y = ax + b$, where a is the gradient of the line and b is the intercept with the y-axis.

11 A control chart is a time-series chart that is used for process control.

12 If the plotted values are within the warning limits the process is under control. If a value is between the warning and action limits another sample is taken. If it is outside the action limits the process is stopped and the machine is reset.

8 Probability

8.1 The meaning of probability

Probability is used to predict the chance of something happening in the future.

There is a chance of snow falling in London on Christmas day.
There is a chance you will win the lottery if you buy a ticket.

A scale of likelihood is shown below.

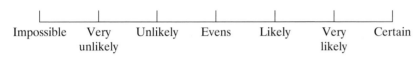

| | | | | | | |
Impossible Very unlikely Unlikely Evens Likely Very likely Certain

Example 1

Write down which of the words on the likelihood scale above are most suitable when describing the following events.

(a) A flower will die.

(b) A baby born will be a boy.

(c) A cow will turn into a horse.

(d) A hurricane will happen in England in the next week.

(a) Certain

(b) Evens

(c) Impossible

(d) Very unlikely

Example 2

Draw a likelihood scale and mark where you think each of the following events should be.

(a) A baby will be born somewhere in the world today.

(b) You will score ten when a normal six-sided dice is rolled.

(c) The jackpot will be won in the next lottery draw.

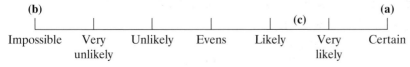

Likelihood of **(b)** < likelihood of **(c)** < likelihood of **(a)**.

In statistics we like to use a numerical scale.

■ **Probability is a numerical measure of the chance of an event happening.**
  – A probability of 0 means it is impossible for the event to happen.
  – A probability of 1 means the event is certain to happen.

A scale using numbers is called a **probability scale**.

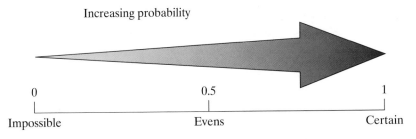

Increasing probability

0   0.5   1

Impossible   Evens   Certain

Example 3

Put the following on a probability scale.

(a) Great Britain will win 50 medals in the next Olympics.

(b) A person chosen at random was born on a weekday.

(c) A die is thrown and does not show a six.

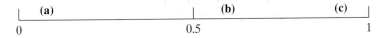

   (a)   **(b)**   **(c)**

0   0.5   1

Probabilities can be written as fractions, decimals or percentages.

Example 4

There is a 20% chance that the Bank of England will change interest rates next month. Write the probability of a change as a percentage, a fraction and a decimal.

$$P(C) = 20\% = \frac{20}{100} = 0.2$$

Stands for probability   Stands for change
(This could be any letter.)

Exercise 8A

1   Write down the words on a likelihood scale that best fit each of the following.

    (a) A tree will die.

    (b) A fair coin will fall tail uppermost.

    (c) There will be a hurricane in England next May.

2 Draw a likelihood scale and mark on it the position of the following.

(a) There will be 4 aces in a pack of 52 cards.

(b) The Queen of England will be president of America.

(c) Every household in Great Britain will make at least one telephone call this week.

3 (a) If something is impossible what is its numerical probability?

(b) If something is certain what is its numerical probability?

(c) Write in words $P(W) = 0.5$.

4 There is an 80% chance of catching a train. What is this probability as a decimal?

5 Draw a probability scale and mark with

(a) a G the probability that the grass will grow this summer,

(b) a D the probability that a rose bush will die,

(c) an O the probability that a dice will show an odd number when thrown.

Odds

Odds are another way of expressing probability. Odds are given as a ratio between the estimated number of failures and the estimated number of successful outcomes.

The ratio failures : successes is the **odds against** an event happening.

The ratio successes : failures is the **odds on** an event happening.

> The odds quoted by bookmakers are not 'true odds' but ones adjusted to give the bookmaker a profit. In the long run you lose and the bookie prospers.

Example 5

A bag contains 6 white balls and 4 red balls. A ball is selected at random. What are the odds of (a) getting a red ball (b) getting a white ball?

There are 2 chances that you will get a red ball for every 3 chances of getting a white ball.

(a) The odds of getting a red ball are 3 to 2 against.

(b) The odds of getting a white ball are 2 to 3 against (or 3 : 2 on).

> If the number of successes is greater than the number of failures, the situation is often called **odds on**.

Odds may be changed to probabilities.

Example 6

If the odds are 7 to 2 against, what is the probability of success?

There are 7 chances of failure to every 2 of success, thus for every $(7 + 2) = 9$ attempts there will be 2 successes and the probability of success is $\frac{2}{9}$.

> The process can be reversed. If the probability of success is $\frac{2}{9}$, for every two successes there are $9 - 2 = 7$ failures. The odds are 7:2 against.

Example 7

A bag contains 3 red, 4 white and 2 blue balls. If a ball is drawn at random what are the odds against:

(a) drawing a red ball?

(b) drawing a blue ball?

(c) drawing a white ball?

(a) There are 3 red balls and $4 + 2 = 6$ non-red balls.
Odds of drawing a red ball are $6 : 3 = 2 : 1$ against.

(b) Odds of drawing a blue ball are $7 : 2$ against.

(c) Odds of drawing a white ball are $5 : 4$ against.

Exercise 8B

1   Describe the odds of 1 to 1 in another way.

2   A bag contains 4 red, 2 white and 3 blue balls. If a ball is drawn at random, work out the odds against it being (a) red, (b) blue, (c) white, (d) red or white.

3   A box contains 20 counters, 7 red, 8 green and the rest yellow.

   (a) How many counters are yellow?

   (b) If a counter is picked at random from the box, work out the odds against it being: (i) green, (ii) yellow, (iii) not red.

4   If the true probability of a horse winning a race is $\frac{2}{5}$ what are the odds against the horse winning?

5   If the true odds of an event being successful are 5 to 2 against, what is the probability of success?

6   A lucky dip contains 3 really good prizes, 10 not so good prizes and 27 poor prizes. Work out the odds against getting:

   (a) a really good prize,

   (b) a poor prize.

   (c) Express the answer to (b) as odds on getting a poor prize.

8.2 Events and outcomes

You may be playing a game where a dice is thrown and a six is required.

The act of throwing the dice is called a **trial**.

The numbers 1, 2, 3, 4, 5 and 6 are all possible **outcomes** of the trial.

■ **The possible results of a trial are called outcomes.**

Being successful and throwing a six is called an **event**.
If any other number appears the event has not occurred.

An event is a set of one or more successful outcomes.

For example, 'throwing a six' is an event that consists of a single outcome, while the event 'throwing a number less than 4' consists of three outcomes with any of the numbers 1, 2 and 3 being accepted as successful.

Equally likely outcomes

If you choose one card from a full pack of 52 cards without any conscious choice then you are equally likely to pick any one of the cards. You have made a **random choice** of the card.

When choosing a card randomly from a pack of 52, each suit is **equally likely** to be an outcome.

When throwing a fair dice all the numbers are equally likely to occur. Throwing a dice gives a random choice from the numbers 1 to 6.

■ **If outcomes have the same chance of happening they are called equally likely outcomes.**

8.3 The probability of an event

If *all* possible outcomes are equally likely then:

■ **Probability of an** = $\dfrac{\text{number of successful outcomes}}{\text{total number of outcomes}}$
event happening

Example 8

A fair six-sided dice is thrown. Work out:

(a) the probability of an even number
(b) the probability of a number $\leqslant 2$
(c) the probability of a multiple of 3.

There are six possible outcomes: 1, 2, 3, 4, 5 and 6. Only three of the six are successful: 2, 4 and 6.

(a) $P(even) = \dfrac{3}{6} = \dfrac{1}{2}$

(b) $P(\leqslant 2) = \dfrac{2}{6} = \dfrac{1}{3}$

1 and 2 are the only successful outcomes.

(c) $P(multiple\ of\ 3) = \dfrac{2}{6} = \dfrac{1}{3}$

3 and 6 are the only multiples of 3 so are the only successful outcomes.

Example 9

This two-way table shows the number of males and females who are right- and left-handed.

	Male	Female	Total
Right-handed	17	20	37
Left-handed	7	6	13
Total	24	26	50

If a person is chosen at random, work out the probability of the following events.

(a) The person is female.

(b) The person is left-handed.

(c) The person is a right-handed male.

There are 50 people altogether.

(a) 26 out of 50 are female so P(*female*) $= \dfrac{26}{50} = \dfrac{13}{25}$

(b) 13 out of 50 are left-handed so P(*left-handed*) $= \dfrac{13}{50}$

(c) 17 out of 50 are right-handed and male so P(*right-handed male*) $= \dfrac{17}{50}$

Exercise 8C

1 List the outcomes for the following:
   (a) The weather tomorrow is studied to see if it rains.
   (b) A fair eight-sided dice, numbered 1 to 8, is thrown.
   (c) A drawing pin is thrown in the air to see how it lands.

2 A fair six-sided dice, numbered 1 to 6, is thrown. Work out:
   (a) the probability of getting a four
   (b) the probability of a number < 5
   (c) the probability that the number is a multiple of 2, but not counting 2 itself.

3 The fair spinner shown is spun. Work out the probability of the arrow pointing to:
   (a) yellow (Y),
   (b) green (G),
   (c) red (R),
   (d) blue (B),
   (e) red or blue (R or B).

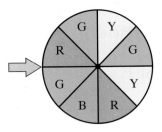

4 Explain what is meant by 'equally likely outcomes'.

5 A card is selected at random from a pack of 52 playing cards. Work out the probability that the card is:
   (a) a heart,
   (b) the King of Hearts,
   (c) a black card,
   (d) a jack.
   (e) Explain why the card had to be picked at random.

6 Six names were written on separate pieces of paper and put into a hat. The names were Ann Smith, Saba Brown, Yves Black, Ann Jones, Jack Brown, and Jack Firth. One name is drawn from the hat. Work out the probability that it was:

(a) someone with the first name Ann,

(b) someone with the first name Yves,

(c) someone with the surname Brown.

7 This two-way table shows the numbers of males and females in a group of 50 who wear or do not wear glasses.

	Male	Female	Total
Wearing glasses	16	18	34
Not wearing glasses	9	7	16
Total	25	25	50

Work out the probability that a person chosen at random is:

(a) female,

(b) not wearing glasses,

(c) a male who wears glasses.

8 Shoppers in a supermarket were asked to taste jams marked A and B. The two-way table shows the preferences of male and female shoppers.

	Male	Female	Total
Jam A	10	13	23
Jam B	2	20	
Total	12		45

(a) Copy and complete the table.

(b) Work out the probability that a shopper chosen at random is:

   (i) a female preferring jam B,

   (ii) a male preferring jam A,

   (iii) a person who prefers jam A.

9 Children going on a boat trip could choose to travel either on the Raven or on the Eagle. The choices of boys and girls are shown in the two-way table.

	Boys	Girls	Total
Raven	11		51
Eagle	35	19	
Total			

(a) Copy and complete the table.

(b) Work out the probability that a child chosen at random is:
- **(i)** female,
- **(ii)** female and chooses Raven,
- **(iii)** chooses Eagle.

10 In a trial of a new drug, eight volunteers are given courses of treatment. In the programme of testing, three volunteers will be given the new drug and the rest will be given an inactive substance. The volunteers are randomly allotted to the treatments. Jean is one of the volunteers. Work out the probability that:

(a) Jean will be given the new drug,

(b) Jean will not be given the new drug.

8.4 Experimental probability

In real life situations the probabilities of different outcomes are not always equal or possible to work out.

In such cases you may need to carry out an experiment or survey to estimate the probability of an event happening.

For example: the probability of a football team winning a game, or the probability of a seed germinating.

■ **The estimated probability that an event might happen** $= \dfrac{\text{number of successful outcomes}}{\text{total number of trials}}$

Example 10

In an experiment a student tosses a coin 100 times and writes down the number of heads after 1, 10, 25, 50, 75 and 100 tosses. The results are shown in the table.

In this case a head is considered a successful outcome of a trial.

Number of throws	1	10	25	50	75	100
Number of heads	1	3	10	28	39	49
Probability of head	1	$\dfrac{3}{10} = 0.3$	$\dfrac{10}{25} = 0.4$	$\dfrac{28}{50} = 0.56$	$\dfrac{39}{75} = 0.52$	$\dfrac{49}{100} = 0.49$

Estimate the probability of getting a head.

A scatter diagram of these results is shown.

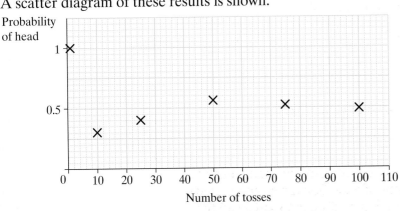

As the number of tosses increases, the closer the probability is to 0.5.

You could try this for yourself.

■ In experiments and surveys, as the number of trials increases, the nearer your estimate for the probability should be to the true value.

Simulation

It may not be possible to carry out an experiment in order to estimate the probability of an event happening. This may be because it is too complex or it is undesirable to carry out the study.

In such cases you can imitate or '**simulate**' the problem.

■ Simulation is the imitation of the conditions of a situation by a theoretical study.

> It would not be sensible to study the spread of an infectious disease by practical experiment.

Example 11

A central warehouse has two loading bays and five lorries. A sixth lorry is being bought. Is another loading bay needed? Explain why simulation would be used to answer this question.

It would be expensive to build the bay if it is not needed and there might be costly delays if the lorry is bought and the bay not built. Several runs of a computer simulation with six lorries and two loading bays would tell the owners if hold-ups are likely to occur. This model could easily be altered and re-run with three bays.

Simulation is:

● quick and cheap
● easily altered
● repeatable.

There are several ways of introducing randomness to a simulation.

● You could use a coin or coins.
● You could use a dice or several dice.
● You could use random numbers.

Example 12

A shop sells ice creams in four different flavours: Vanilla, Chocolate, Raspberry and Strawberry. Over a long period of time:

3 out of ten customers choose Vanilla,
2 choose Chocolate,
1 chooses Raspberry,
4 choose Strawberry.

The shop has an average of 20 customers each day.

(a) Run a simulation to show a typical day.
(b) Repeat your simulation.
(c) Explain how repeat simulations could help the shopkeeper to estimate the minimum amount of each flavour he should have in stock at the beginning of each day, if he is to satisfy all the possible demands of his customers.

Generate 20 random numbers on your calculator. Multiply each number by 10 and add on 1. Take the whole number (this will be a number from 1 to 10).

> If the calculator gives the random number 0.962 then $10 \times 0.962 + 1 = 10.62$. The random numer is 10.

(a) Allocate numbers as follows:
   1, 2 and 3 = Vanilla
   4 and 5 = Chocolate
   6 = Raspberry
   7, 8, 9 and 10 = Strawberry.

Random number	9	7	2	4	9	5	2	1	1	8	4	1	4	10	6	10	8	9	3	5
Flavour	S	S	V	C	S	C	V	V	V	S	C	V	C	S	R	S	S	S	V	C

This run gives 8 Strawberry, 6 Vanilla, 5 Chocolate and 1 Raspberry.

(b)

Random number	7	1	9	8	5	9	2	3	10	10	5	1	10	5	7	10	2	4	4	2
Flavour	S	V	S	S	C	S	V	V	S	S	C	V	S	C	S	S	V	C	C	V

The second run gives 9 Strawberry, 6 Vanilla, 5 Chocolate and 0 Raspberry.

(c) From the two runs the maximum sales of each variety are 9 Strawberry, 6 Vanilla, 5 Chocolate and 1 Raspberry. If the simulation is repeated many times the shopkeeper will have a better idea as to the maximum daily demand for each flavour.

Using ICT for simulation

If you have tried the experiment of tossing a coin to find the probability of getting a head, you will have found that it needs a large number of tosses. This takes a long time.

You could do this instead by using ICT to give a simulation.

You could generate random numbers between 0 and 1. If a number is less than 0.5 call it a 'head', if greater than 0.5 call it a 'tail'.

Example 13

(a) Use a spreadsheet to simulate the tossing of a coin 20 times.

(b) Draw a graph of the results.

(c) What is the probability of getting a head?

(a) In this program the formula '=IF(A3<0.5, "Heads", "Tails")' in cell B3 means if the number in cell A3 is less than 0.5 put "Head" in cell B3, otherwise put "Tail". To produce the spreadsheet follow these instructions.

- Open a new Excel document.
- Enter headings in row 1 as shown below.

	A	B	C	D	E	F
1	Rand No.	Result of toss	Total No. of Heads	Total No. of tosses	Proportion of Heads	
2						
3						

- In cell A3 enter '=RAND()'. Click on ✓. (This generates a random number between 0 and 1 in A3.)
- In cell B3 enter '=IF(A3<0.5, "Heads", "Tails")'. Click on ✓. (The cell shows the result of the first toss.)
- In cell C3 enter '=IF(B3="Heads", C2+1,C2)'. Click on ✓. (This totals up the number of heads.)
- In cell D3 enter '=(D2+1)'. Click on ✓. (This totals the number of tosses.)
- In cell E3 enter '=C3/D3'. (This works out the proportion of heads.)
- Highlight cells A3 to E3. Place the cursor on bottom right hand corner of cell E3, hold down the left mouse button and drag down to cell E22. (This gives you a further 19 tosses.)

(b) To draw a graph of the results:
- Highlight cells D3 to E22, and click the **chart wizard** button.
- From chart types select **XY(Scatter)**.
- Click **Next** until you get to **Chart options**.
  In **Value(X) axis** box type 'Number of tosses'.
  In **Value(Y) axis** box type 'Proportion of heads'.
- Click **Next**.
- Select **New Sheet** option.
- You can now print the graph.

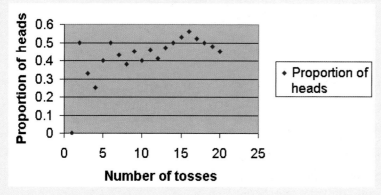

This graph shows the result of a first run.

Because the random numbers change each time, your chart will be different to this one.

● Click on **Sheet 1**. Click on **B3**, re-enter 0.5 in the formula then click another square. You will get a re-run of the tossing experiment.

● Click on **Chart 1** to see the new graph.

(c) In this case the probability of a head at the end of the run happened to be 0.5 (which is the actual true value for a fair coin). Your answer might be different to this. The result of a second run might have been different again, but, if you do enough runs noting the final value each time, you will get a better estimate of the true probability.

Exercise 8D

1 In an experiment 50 poppy seeds are set and 30 produce plants. Work out an estimate for the probability of a poppy seed producing a plant.

2 In a nursery school a class of 20 children each plant a bean seed in a jam jar. 17 of the bean seeds grow into bean plants. Work out an estimate for the probability that a bean seed will grow into a plant.

3 Under what circumstances would you use a simulation rather than an experiment?

4 In a works canteen the four meals offered are a salad, a roast, pasta or a vegetarian meal. Over a period of time it was found that out of every 10 workers:
2 chose salad,
5 chose roast,
1 chose vegetarian,
2 chose pasta.
Each lunchtime the canteen serves 30 workers.

(a) Run a simulation using random numbers on your calculator.

(b) Repeat your simulation.

(c) Explain how repeated simulations could help the chef to decide how many of each type of meal he should produce.

5 A country park wants to make sufficient provision for wet weather activities. It is going to create four activity centres. A survey of visitors shows that out of every 10 people:
5 visitors would choose centre A,
1 would choose centre B,

2 would choose centre C and
2 would choose centre D.
The park normally has 50 visitors on a rainy day.

   (a) Use a calculator or random number tables to generate 50
       random numbers from 1 to 10.

   (b) Suggest how you would use these numbers to simulate a
       rainy day with 50 visitors.

   (c) Why would it be sensible for the park to run several
       simulations?

6   Using the formula =INT(1+6*RAND()) (which gives
    random numbers from 1 to 6), and a spreadsheet, simulate
    20 tosses of a dice.

7   Use Excel to simulate 30 throws of a dice with 8 faces on
    which are the numbers 1 to 8.

8   Use Excel to simulate 20 spins of a triangular spinner with
    equal sections coloured red, green and yellow. Get your
    spreadsheet to count up how many times each colour appears.

9   A game for two players has the following rules:
    (i) Each player puts 50p in the kitty.
    (ii) Each player takes turns at throwing a fair dice and gets
         20p if they throw a 6. If they do not throw a 6 they put
         1p in the kitty. Use an Excel simulation to
    (a) run the game until each player has had 10 tosses,
    (b) draw a graph of how much is left in the kitty after each
        throw.

Using probability to assess risk

What is the risk of your house burning down this year?

If you look at the number of houses like yours that are burnt down
each year and compare it to the total number of houses like yours,
you can estimate the probability of your house burning down. If you
count your house burning down as an event (successful outcome):

Estimated probability of your house burning down

$$= \frac{\text{number of successful outcomes}}{\text{total number of trials}} = \frac{\text{number of houses burnt}}{\text{total number of houses}}.$$

Insurance companies will use this information to set your insurance
premiums. This is called **risk assessment**.

If the risk of your house burning down in any year is p, then the
insurance company should charge you £1000 $\times p$ for each £1000 of
house value (i.e. $p \times$ house value).

(Probably it will be a bit more than this to allow for running costs and profits.)

Cost of risk insurance = money at risk × risk assessment.

Example 14

Using past records, an insurance company assesses the yearly risk of a house in a certain area being flooded. During the last 50 years flooding in that area has occurred only once. It uses this risk assessment to work out the cost of the flood risk part of its annual premiums.

(a) What is the risk assessment?
(b) If the average pay-out when a house is flooded is £1200, what would you expect to pay for insurance against being flooded?

(a) Risk assessment = $\dfrac{\text{number of successful outcomes}}{\text{total number of trials}} = \dfrac{1}{50} = 0.02$

(b) Cost of flood insurance = money at risk × risk assessment
$$= £1200 × 0.02 = £24$$

Exercise 8E

1   John is 18 and has just passed his driving test. An insurance company finds that young men of John's age and living in John's postal district had 4 accidents in the past year. There were 150 drivers like John in the area during the last year. What is the risk assessment for John having an accident this year?

2   An insurance company that insures all types of boats has found that out of 800 boats insured 20 had a shipping accident of some sort in the past year. What risk assessment would you make of a particular boat having an accident?

3   The risk assessment of a particular type of company going bankrupt during a one-year period is $\frac{2}{27}$. If a small country has 54 companies of this type how many are likely to end up bankrupt this year?

4   A bank finds that in the past, 1.5% of its customers have been unable to pay their mortgage payments each year. The bank has an insurance policy to cover this eventuality. If a customer has mortgage repayments of £3200, what is a fair yearly charge for this customer's mortgage repayments insurance policy?

5   When going on holiday abroad you are told to take out insurance to cover the cost of possible medical treatment. The average cost of medical treatment abroad last year was £5000, and 2% of people going abroad made claims. Using these figures, how much would you expect to pay for this insurance if you do not include the insurance company's expenses and profits?

8.5 Sample space

To help find the probability of one, two or more events occurring you can list all the possible outcomes.

> For example, if you roll a dice there are six possible outcomes. The sample space is S = (1, 2, 3, 4, 5, 6).

■ **A list of all possible outcomes is called a sample space.**

If there are two events, a table can represent the sample space.

Example 15

A fair coin is tossed and a fair dice is thrown.

(a) Draw up a table of all possible outcomes.
(b) Work out the probability of getting a head and throwing a six.
(c) Work out the probability of a head and an even number.

(a)

		Dice					
		1	**2**	**3**	**4**	**5**	**6**
Coin	**Head (H)**	H, 1	H, 2	H, 3	H, 4	H, 5	H, 6
	Tail (T)	T, 1	T, 2	T, 3	T, 4	T, 5	T, 6

(b) $P(H, 6) = \dfrac{1}{12}$

> There are 12 equally likely outcomes of which one is a head and a six.

(c) $P(H, even) = \dfrac{3}{12} = \dfrac{1}{4}$

> There are 12 equally likely outcomes of which 3 are a head and an even number.

Example 16

Two fair dice are thrown and the scores added together.

(a) Draw a sample space diagram showing all the possible outcomes.
(b) Work out the probability of getting a total of 10.
(c) Work out the probability of getting a total that is even.

(a)

		Dice 1					
		1	**2**	**3**	**4**	**5**	**6**
Dice 2	**1**	2	3	4	5	6	7
	2	3	4	5	6	7	8
	3	4	5	6	7	8	9
	4	5	6	7	8	9	10
	5	6	7	8	9	10	11
	6	7	8	9	10	11	12

(b) $P(10) = \dfrac{3}{36} = \dfrac{1}{12}$

> There are 36 equally likely outcomes of which 3 are successful.

(c) $P(even) = \dfrac{18}{36} = \dfrac{1}{2}$

> There are 36 equally likely outcomes of which 18 are even numbers.

In simple cases, a sample space can be used to find the probability of three events occurring.

Example 17

Three fair coins are tossed.

(a) Write down the sample space.

(b) Work out the probability of 3 heads.

(c) Work out the probability of exactly 2 heads.

(a) S =

HHH		
HHT	*HTH*	*THH*
HTT	*THT*	*TTH*
TTT		

There are 8 equally likely outcomes.
Note *HHT* is a different outcome to *HTH*.

(b) P(*3 heads*) $= \dfrac{1}{8}$

There are 8 equally likely outcomes of which 1 is 3 heads.

(c) P(*2 heads*) $= \dfrac{3}{8}$

There are 8 equally likely outcomes of which 3 contain 2 heads.

Exercise 8F

1   Two fair coins are tossed.
    (a) Draw a table to find the sample space and list on it all possible outcomes.
    (b) Work out the probability of getting a head and a tail.
    (c) Work out the probability of getting two heads.

2   Two dice, each having sides numbered from 1 to 6, are tossed and the total of the scores noted.
    (a) Draw a table to find the sample space and on it put the total scores.
    (b) Write down the probability of a total score of 12.
    (c) Write down the probability of a total score ≤ 7.
    (d) Write down the probability of a total score > 10.
    (e) Write down the probability of a total score that is even.

3   A team from a tennis club consists of 4 men players *A*, *B*, *C* and *D*, and two women players, *X* and *Y*. To decide who will represent the club in mixed doubles, a fair spinner lettered *A*, *B*, *C* and *D* to represent the men is spun, and a fair coin to represent the two women players is tossed, a head representing *X* and a tail representing *Y*.

(a) Draw up a table to find the sample space.

(b) Write down the probability that the man selected is *D*.

(c) Write down the probability that *X* and *B* form a pair for mixed doubles.

(d) Write down the probability that *Y* and either *B*, *C* or *D* form a pair for mixed doubles.

4 A train travelling from London to Swindon via Reading is equally likely to take 1, 2, 3, 4, 5 or 6 minutes longer than the timetable allows between London and Reading, and is equally likely to take 1, 2, 3 or 4 minutes longer than the timetable allows between Reading and Swindon. Assume that trains never arrive on time or earlier than the time tabled, and are never later than the times shown above.

(a) Draw a table to find the sample space.

(b) Write down the probability that a train is 8 minutes late arriving at Swindon.

(c) Write down the probability that it is less than 3 minutes late arriving at Swindon.

(d) Write down the probability that it is $\geqslant 6$ minutes late arriving at Swindon.

5 Pens are equally likely to be Blue, Red or Green. Pen caps are equally likely to be Blue, Red or Green. Caps are assembled to the pens randomly.

(a) Draw a space diagram to show all possible outcomes.

(b) Work out P(*a red pen has a red cap*).

(c) Work out P(*a pen has a matching cap*).

(d) Work out P(*a blue pen has a green cap*).

6 The total score on the three spinners shown is used for a board game.

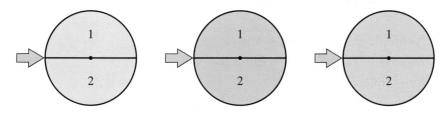

(a) Write down the sample space for the outcomes.

(b) Calculate the probability of a total score of 6.

(c) Calculate the probability of a total score of 2.

(d) Calculate the probability of a total score of 4.

(e) Calculate the probability of a total score < 6.

7   Ann, Brenda and Carol go shopping together in a busy supermarket.
    They each join a different queue. All the queues are the same length.

   (a) Write down the sample space for the order in which they leave the supermarket.

   (b) Work out the probability that Ann leaves before Carol.

   (c) Work out the probability that Brenda is the last to leave.

8.6 Venn diagrams

■   **A Venn diagram is a diagram representing a sample space.**

A **Venn diagram** may be used to calculate probabilities.

Each region of a Venn diagram represents a different set of data.

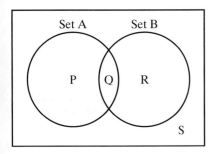

Area P represents the data that are in set A but not in set B.
Area R represents the data that are in set B but not in set A.
Area Q represents the data that are in set A and also in set B.
Area S represents the data that are not in set A and not in set B.

Example 18

In a study on the effects of drugs there are 70 patients. 24 receive drug B, 30 receive drug A and 20 receive both drug A and drug B.

(a) Draw a Venn diagram to represent these data.

(b) Calculate the probability that a patient chosen at random is receiving both drugs A and B.

(a) First draw and label a Venn diagram, and put in the numbers you know.

   The only one you can write in immediately is 20.

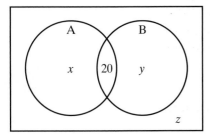

There are 30 in group A, so

$$x + 20 = 30$$

There are 24 in group B, so

$$y + 20 = 24$$

Therefore $x = 10$ and $y = 4$.

The total number must be 70, so

$$z = 70 - 10 - 20 - 4 = 36$$

The final Venn diagram is

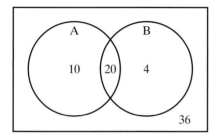

(b) P(*neither drug A nor drug B*) $= \dfrac{36}{70} = \dfrac{18}{35}$

There are 70 patients and 36 take neither drug.

Rather than using numbers in a Venn diagram you can enter probabilities.

If probabilities are entered instead of numbers the diagram in Example 18 looks like this:

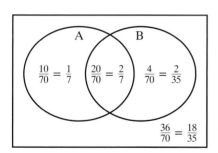

Example 19

The prime minister makes a speech. Two newspapers differ in their editorial policies. The probability that paper G will put the speech as their main headline is 0.2. The probability that paper L will put the speech as their main headline is 0.8. The probability that neither put the prime minister's speech as their main headline is 0.1.

Draw a Venn diagram to represent these probabilities and use it to find the probability that both use it as their main headline.

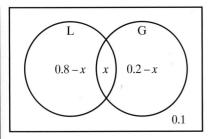

Letting the unknown probability that both use it as their main headline be x, you can then find expressions for the other areas.

Since the probabilities must add up to 1:

$$(0.8 - x) + x + (0.2 - x) + 0.1 = 1$$
$$1 - x + x - x + 0.1 = 1$$
$$1.1 - x = 1$$
$$x = 0.1$$

The probability of both newspapers using the prime minister's speech as their main headline is 0.1.

Example 20

In an office there are 30 people; 12 have 'A' levels in Art (A), 8 have 'A' levels in Biology (B), 8 have 'A' levels in Latin (L), 3 have 'A' levels in Art and Biology, 3 have 'A' levels in Biology and Latin, 4 have 'A' levels in Latin and Art and 2 have 'A' levels in Art, Biology and Latin.

(a) Draw a Venn diagram to represent these data.

(b) One person is chosen at random. Calculate the probability that they have 'A' levels in:
  (i) at least one of the three subjects
  (ii) only one of the three subjects
  (iii) Latin but not Biology.

(a) This time you have three subjects so you must draw three circles. Start by filling in the number 2 in the intersection of the three circles. Then work outwards.

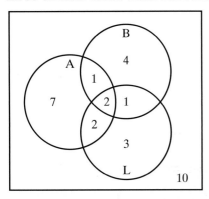

Two people have all three subjects.

(b) **(i)** There are 30 people altogether, of whom 10 have 'A' levels in none of the three subjects. Therefore, 20 have 'A' levels in at least one subject.

P('A' levels in at least one of the three subjects) $= \dfrac{20}{30} = \dfrac{2}{3}$.

(ii) 7 have 'A' levels only in Art, 4 have 'A' levels only in Biology, 3 have 'A' levels only in Latin.

$$P(\text{'A' levels in only 1 subject}) = \frac{14}{30} = \frac{7}{15}.$$

(iii) 5 study Latin but not Biology.

$$P(\text{'A' levels in Latin but not Biology}) = \frac{5}{30} = \frac{1}{6}.$$

Exercise 8G

1   The following Venn diagrams were drawn to represent school subjects taken by a group of 50 children. Work out the value of x in each case and say what x represents.

(a)

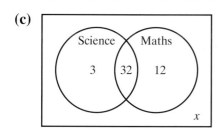

(b)

(c)

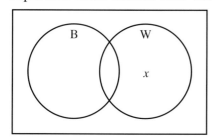

2   The following Venn diagram represents the fur colour of 60 rabbits. Let B represent the rabbits with 'some black fur'. Let W represent those rabbits with 'some white fur'.

Of the rabbits, 6 have only black fur, 20 have black and white fur, and 16 have no black or white fur.

(a)   Copy and complete the diagram.
(b)   Work out the value of x.
(c)   How many rabbits have some black fur?
(d)   What is the probability that a rabbit picked at random has no black or white fur?

3 Copy the following Venn diagrams, filling in the missing
   probabilities.

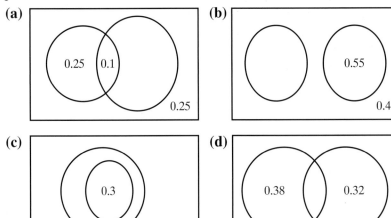

(a) 0.25 0.1 0.25

(b) 0.55 0.4

Remember: all the
probabilities added
together must equal 1.

(c) 0.3 0.2

(d) 0.38 0.32 0.21

4 Out of 120 people questioned, 60 said they had taken aspirin when
   they had a headache, 28 had taken codeine and 24 had taken both.
   (a) Draw a Venn diagram to represent these data.
   (b) Work out the probability that a person picked at random
   from the group takes neither aspirin nor codeine.

5 A town council gets a grant to build two sports centres. During
   the first year the centres were open it was found that 36% of the
   population had been to centre A, 22% had been to centre B and
   10% had been to both.
   (a) Draw a Venn diagram to represent these data.
   (b) What percentage of the population had been to neither centre?
   (c) What percentage had been to centre A only?

6 In a child nursery there are 30 children. 15 children like orange
   squash, 10 children like both orange squash and milk and 8
   children like neither orange squash nor milk.
   (a) Draw a Venn diagram to represent these data.
   (b) How many children like milk?
   (c) Work out the probability that a child chosen at random
   likes milk but dislikes orange squash.

7 In an aviary containing 30 parrots the probability of a parrot
   having only green feathers is 0.3, and the probability of a parrot
   having both red and green feathers is 0.1. The probability of a
   parrot having no green or red feathers is 0.45.
   (a) Draw a Venn diagram to represent these probabilities.
   (b) Calculate the probability that a parrot chosen at random
   will have some red feathers.
   (c) How many parrots will have green feathers and no red
   feathers?

8 A music school has 100 students. 50 students play the piano, 20 play the violin, and 60 play the oboe. There are 10 students who play all three instruments, and 22 who play two (15 who play both the piano and the oboe, 5 who play both the piano and violin, and 2 who play both the violin and the oboe).

   (a) Copy the Venn diagram below and complete it by putting in appropriate numbers.

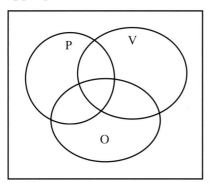

   (b) Work out the probability that a student chosen at random does not play any of the three instruments.

   (c) Work out the probability that the student plays only the piano.

9 The Venn diagram shows the probability of events A, B and C happening.

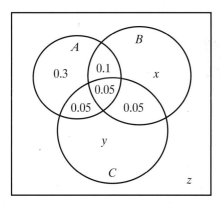

   Given that $P(B) = 0.4$ and $P(C) = 0.35$, work out x, y and z.

10 A market researcher asked 100 schoolchildren which of three different chocolate bars they liked. Of the 100 children, 25 liked bar A, 30 liked bar B and 20 liked bar C. There were 6 who liked both bars A and B, 7 liked both bars B and C, 5 liked both bars C and A and 40 who liked none. Three children liked all three bars.

   (a) Draw a Venn diagram for these data.

   (b) A child is selected at random. Work out:

   (i) P(*the child liked at least one of the bars*)

   (ii) P(*the child liked only one of the bars*)

   (iii) P(*the child liked none of the bars*).

8.7 Mutually exclusive outcomes

When a dice is thrown the outcome is either even or odd. It is not possible to have an odd number and an even number showing at the same time. The event in which an odd number occurs and the event in which an even number occurs are **mutually exclusive**.

■ **Outcomes are mutually exclusive if they cannot happen at the same time.**

Example 21

Which of the following are mutually exclusive events?

(a) You get the number 2 and an odd number on a single throw of a dice.

(b) The next car you see is red and three years old.

(c) The sun will shine and the temperature will be below freezing.

Only the events in **(a)** are mutually exclusive.

It is possible to get a 3-year-old red car.
It is possible for the sun to shine when it is freezing.

The addition law for mutually exclusive events

■ **For two mutually exclusive events A and B**

$$P(A \text{ or } B) = P(A) + P(B)$$

This is called the **addition law** for mutually exclusive events.

It is sometimes called the **OR rule**.

Example 22

A card is drawn from a pack of 52 cards. Work out the probability that the card drawn is either an ace or a king.

There are 52 cards and 4 are aces so $P(ace) = \dfrac{4}{52} = \dfrac{1}{13}$

There are 4 kings so $P(king) = \dfrac{4}{52} = \dfrac{1}{13}$

$P(ace \text{ or } king) = P(ace) + P(king) = \dfrac{1}{13} + \dfrac{1}{13} = \dfrac{2}{13}$

Note that this law can be extended for three or more mutually exclusive events. You just add the probabilities. For example,

$$P(A \text{ or } B \text{ or } C) = P(A) + P(B) + P(C)$$

Example 23

In cricket, if the batsman hits the ball over the boundary he scores 4 runs. If he hits it over the boundary without it bouncing he gets 6 runs. If the batsman does not hit the ball over the boundary he can score runs by running between the wickets.

It is the last ball of a cricket match. Jack's side need 4 runs to win. The probability of Jack scoring 4 runs by getting a boundary is 0.08. The probability that he hits the ball over the boundary and gets 6 runs is 0.03. The probability that Jack gets 4 runs by running between the wickets is 0.01.

What is the probability of Jack:

(a) getting a 4 or a 6 by hitting the ball over the boundary?

(b) making a score of 4?

(c) winning the match for his team?

(a) P(*hitting a 4 or 6*) = P(*hitting a 4*) + P(*hitting a 6*)

$$= 0.08 + 0.03$$
$$= 0.11$$

(b) P(4) = P(*running 4*) + P(*hitting a 4*)

$$= 0.01 + 0.08$$
$$= 0.09$$

(c) P(*winning*) = P(*hitting a 4*) + P(*hitting a 6*) + P(*running 4*)

$$= 0.08 + 0.03 + 0.01$$
$$= 0.12$$

8.8 Exhaustive events

■ **A set of events is exhaustive if the set contains all possible outcomes.**

Example 24

A fair dice is thrown.

A is the event 'a score $\leqslant 3$'.
B is the event 'a score > 3'.
C is the event 'an even number'.

Decide whether the following pairs of events are exhaustive:

(a) A and C

(b) A and B

(c) B and C.

(a) A and C are not exhaustive as 5 is not included.

(b) A and B are exhaustive as all the numbers are included.

(c) B and C are not exhaustive as 1 and 3 are not included.

8.9 The sum of the probabilities for a set of mutually exclusive, exhaustive events

■ **For a set of mutually exclusive, exhaustive events $\Sigma p = 1$.**

For a fair die the possible outcomes are 1, 2, 3, 4, 5 and 6. Each outcome is equally likely and therefore the probability of each outcome is $\frac{1}{6}$. We use P for the word 'probability', so the probability of a 1 is written as P(1).

$$P(1) + P(2) + P(3) + P(4) + P(5) + P(6) = \tfrac{1}{6} + \tfrac{1}{6} + \tfrac{1}{6} + \tfrac{1}{6} + \tfrac{1}{6} + \tfrac{1}{6} = 1$$

> We use p for a particular probability, so Σp means the sum of all the individual probabilities.

> 1, 2, 3, 4, 5 and 6 are all possible outcomes. They are mutually exclusive since only one can happen at a time.

The probability of an outcome not happening

The probability of an event A not happening is written as P(not A).

Since an event either happens or does not happen, $\Sigma p = 1$.

$$P(A) + P(\text{not }A) = 1$$

so:

■ **for an event A, P(not A) = 1 − P(A).**

> $P(\overline{A})$ is an alternative.

Example 25

Petra's father would be happy if she were to marry either a millionaire or, if not a millionaire, a man who is likely to become one in the next 5 years. From an annual digest of statistics he finds that if A is the event 'Petra marries a millionaire' then P(A) = 0.02 and if B is the event 'Petra marries a man who is likely to become a millionaire in the next 5 years' then P(B) = 0.08.

(a) What is the probability of her father being happy?

(b) What is the probability of her making her father unhappy?

(a) Let C be the event 'Petra makes her father happy'.

$$P(C) = P(A \text{ or } B) = P(A) + P(B) = 0.02 + 0.08 = 0.1$$

(b) P(not C) = 1 − P(C) = 1 − 0.1 = 0.9

Exercise 8H

1. Which of the following are mutually exclusive events?
   (a) A drawing pin falling head down and a drawing pin falling point down.
   (b) A student studying Mathematics and studying French.
   (c) Getting both a six and a one when throwing a dice once.

2. If A, B and C are mutually exclusive events and P(A) = 0.2, P(B) = 0.4 and P(C) = 0.3, work out:
   (a) P(A or B), (b) P(A or C), (c) P(C or B).

3 Which of the following events are exhaustive?

   (a) The event 'getting a head' and the event 'getting a tail' when tossing a coin.

   (b) The event 'getting two heads' and the event 'getting a head and a tail' when tossing a fair coin twice.

   (c) The event 'winning a game of darts' and the event 'losing a game of darts'.

4 The fair spinner shown is spun.
The event X is a score > 4.
The event Y is an even number.
The event Z is a score $\leqslant 4$.

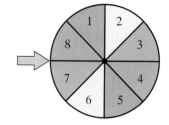

   Which of the following are exhaustive events?

   (a) X and Y

   (b) X and Z

   (c) Y and Z

5 A random number table is used to give numbers between 1 and 8.

   (a) Work out the probability of getting a 1.

   (b) Work out the probability of getting a 2.

   (c) Work out the probability of getting any of the numbers 1 to 8.

6 A coin is biased in such a way that the probability of getting a head is 0.6. Calculate the probability of not getting a head.

7 In a class of 30 children, 10 have a cooked breakfast only, 5 have a cereal only and the rest have just toast. Let A be the event 'have a cooked breakfast', B the event 'cereal for breakfast', and C the event 'toast for breakfast'. Work out:

   (a) how many children have just toast,

   (b) the probability that a child has a cooked breakfast,

   (c) the probability that a child has just cereal,

   (d) $P(A \text{ or } B)$,

   (e) $P(A \text{ or } C)$,

   (f) $P(\text{not } A)$.

8 The probability of having black hair is 0.6, the probability of having red hair is 0.2 and the probability of having blonde hair is 0.15. These events are mutually exclusive.

   (a) Work out the probability of having either black or blonde hair.

   (b) Work out the probability of having black, red or blonde hair.

   (c) Work out the probability of not having red hair.

8.10 The multiplication law for independent events

■ **Two events are independent if the outcome of one event does not affect the outcome of the other event.**

If two events A and B are independent the probability that they both happen is written P(A and B).

■ **For two independent events A and B**

 P(A and B) = P(A) × P(B)

This is called the **multiplication rule** for independent events.

Example 26

Helen likes cuddly toys. The probability that her grandmother buys her a cuddly toy for her birthday is 0.7. The probability that her mother buys her a cuddly toy is 0.5. Assuming that her mother and grandmother do not discuss what they are each going to buy her for her birthday, work out the probability that she gets a cuddly toy from her mother and from her grandmother.

The two events are independent.

If G is the event that 'she gets a cuddly toy from her grandmother' and M is the event that 'she gets a cuddly toy from her mother' then:

 P(G) = 0.7 and P(M) = 0.5

 P(G and M) = P(G) × P(M)
  = 0.7 × 0.5
  = 0.35

Note that this law can be extended for three or more independent events. You just multiply the probabilities. For example:

 P(A and B and C) = P(A) × P(B) × P(C)

Example 27

An office has bought three computers. The probability of a computer breaking down in the first year is 0.03. What is the probability that all three break down in the first year?

The three events are independent.

Let A be the event the first computer breaks down in the first year.
Let B be the event the second computer breaks down in the first year.
Let C be the event the third computer breaks down in the first year.

 P(A and B and C) = P(A) × P(B) × P(C)
  = 0.03 × 0.03 × 0.03
  = 0.000 027

When a coin is tossed and a dice is rolled the result on the coin does not affect the outcome for the dice.

The events must be independent to use this rule.

If two events are independent, they are not mutually exclusive and:

 P(A or B) = P(A) + P(B)
  − P(A and B)

This is because

 P(A and B) is included in both P(A) and P(B).

P(A and B) is subtracted so that it is only included once.

In Example 26

 P(G or M) = P(G) + P(M)
  − P(G and M)
  = 0.7 + 0.5 − 0.35
  = 0.85

Exercise 8I

1   Which of the following are independent events?
   (a) Throwing a six with a dice and picking an ace from a pack of cards.
   (b) Having red hair and a letter C in your name.
   (c) Being male and being bald.

2   If A, B and C are independent events, and $P(A) = 0.3$, $P(B) = 0.2$ and $P(C) = 0.4$, work out:
   (a) $P(A$ and $B)$, (b) $P(B$ and $C)$, (c) $P(A$ and $C)$, (d) $P(A$ or $C)$.

3   The probability of a student taking a packed lunch to school is 0.7. The probability of a student walking to school is 0.6.
   (a) Are these events independent?
   (b) What is the probability of a student walking to school and taking a packed lunch?

4   The probability of Yoko going sailing on a Tuesday is $\frac{1}{7}$. The probability that she will have pasta for dinner on a Tuesday is $\frac{4}{5}$. These are independent events. Calculate:
   (a) the probability that she will go sailing and have pasta for dinner on a Tuesday,
   (b) the probability that she will not sail on a Tuesday,
   (c) the probability that she will not sail and not have pasta on a Tuesday.

5   The probability of a football player scoring a goal is $\frac{1}{12}$. The probability that he will be injured in a match is $\frac{1}{40}$. The probability of his team winning is $\frac{2}{3}$. These events are independent. Calculate the probability that:
   (a) he scores and is injured,
   (b) he scores and his team wins.

6   An alarm system has a stand-by battery that keeps the system working when the main electrical supply breaks down. The probability of a supply failure in any given week is 0.04. The probability of a battery failing in any given week is 0.15. What is the probability of both failing in any given week?

7   The probability of a woman having toast for breakfast is 0.3. The probability of her newspaper being delivered on time is 0.75. The probability that she will go to work in her car is 0.2. Assuming these are independent, what is the probability that:
   (a) she will have toast for breakfast and her newspaper will be delivered on time,
   (b) she will not go to work by car,
   (c) she will have toast for breakfast and not go to work by car.

Tree diagrams

For combined events a **tree diagram** can be used.

■ **A tree diagram can be used to aid calculation.**

Each branch of the tree represents an outcome with the probability of the outcome being written on the branch.

Example 28

A bag contains 5 red balls (R) and 4 green balls (G). A ball is chosen at random, the colour is noted and the ball is then replaced in the bag. A second ball is then chosen and the colour noted. Draw a tree diagram to represent this information.

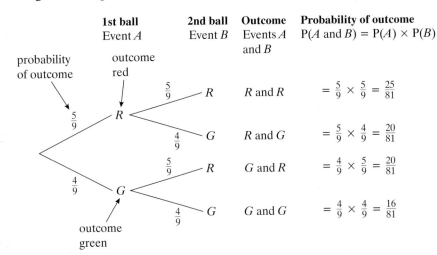

Note the sum of the probabilities of the outcomes is 1.
$$\tfrac{25}{81} + \tfrac{20}{81} + \tfrac{20}{81} + \tfrac{16}{81} = \tfrac{81}{81} = 1$$

Each path through the tree from left to right produces a different outcome.

The probability of each outcome is found by multiplying together the probabilities on the branches of the path taken.

The addition law can be used by adding together appropriate outcomes at the ends of the branches.

For example:

P(*ending up with one of each colour*) = P(R and G) + P(G and R)
$$= \tfrac{20}{81} + \tfrac{20}{81}$$
$$= \tfrac{40}{81}$$

Example 29

A company is to employ three new recruits and it interviews equal numbers of males and females. Because of the company's equal opportunities policy the recruits are equally likely to be male or female. Draw a tree diagram and find from it the probability of all three recruits being female.

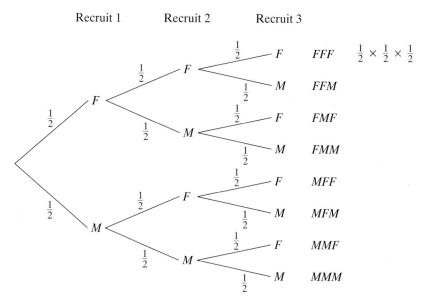

$$P(FFF) = \tfrac{1}{2} \times \tfrac{1}{2} \times \tfrac{1}{2} = \tfrac{1}{8}$$

Sometimes one branch of a tree may stop earlier than the others.

Example 30

A mechanic checks four identical machines to see which one has broken down. Each machine is equally likely to be the broken one. He checks each machine in turn until he finds the one that has broken down.

(a) Draw a tree diagram to represent this information.

(b) Work out the probability of:
   (i) the broken machine being the first he looks at,
   (ii) the broken machine being the last one he looks at.

(a) The probability that the machine he is checking is the broken one is $\tfrac{1}{4}$.

The probability it has not broken down is $1 - \tfrac{1}{4} = \tfrac{3}{4}$.

1st machine	2nd machine	3rd machine	Outcome	Probability of outcome
$\tfrac{1}{4}$ B			B	$= \tfrac{1}{4}$
$\tfrac{3}{4}$ N	$\tfrac{1}{4}$ B		NB	$= \tfrac{3}{4} \times \tfrac{1}{4} = \tfrac{3}{16}$
	$\tfrac{3}{4}$ N	$\tfrac{1}{4}$ B	NNB	$= \tfrac{3}{4} \times \tfrac{3}{4} \times \tfrac{1}{4} = \tfrac{9}{64}$
		$\tfrac{3}{4}$ N	NNN	$= \tfrac{3}{4} \times \tfrac{3}{4} \times \tfrac{3}{4} = \tfrac{27}{64}$

If the first three are not broken the fourth must be the broken one.

(b) **(i)** P(1st machine broken) = P(B) = $\tfrac{1}{4}$
   (ii) P(*broken machine is the last one looked at*) = P(NNN) = $\tfrac{27}{64}$

Exercise 8J

1   A boy picks a card from a pack of 52 cards and a girl tosses a fair coin.

   (a) Work out the probability that the boy picks a red card.

   (b) Work out the probability that the girl gets a head.

   (c) Copy and complete the tree diagram.

Card colour	Result of toss	Outcome	Probability of outcome

   ```
                                 H        B H
                        B
               0.5            T        B T

                                 H
                        R
                             T
   ```

   (d) What is the probability of the boy getting a red card and the girl getting a tail?

2   A box contains 3 red beads and 4 green beads. A bead is drawn from the box and its colour is noted. The bead is returned to the box and after the box is shaken a second bead is drawn.

   (a) Draw a tree diagram to show all the different outcomes for the two bead colours.

   (b) Work out the probability of getting two red beads.

   (c) Work out the probability of getting one red and one green bead.

3   A government-financed research unit randomly selected a factory for study from a list of 20 factories. Eight of the factories were in the north of the country and the rest in the south. Six of the northern factories do heavy engineering, the rest do light engineering. Six of the factories in the south do heavy engineering.

   (a) Write down the probability that a factory in the north will do heavy engineering.

   (b) Write down the probability that a factory in the south will be doing heavy engineering.

   (c) Draw a tree diagram.

   (d) Work out the probability that a factory chosen at random will be doing heavy engineering in the north of the country.

   (e) Work out the probability that the factory chosen will be doing light engineering.

4 In a survey of 60 women and 40 men, 60% of each sex said that they smoked cigarettes. A person chosen at random is sent a follow-up questionnaire. Draw a tree diagram to show the probability of the person chosen being a man/woman who smokes/does not smoke.

5 The following tree diagram shows the events:
It will rain tomorrow (R), it will not rain tomorrow (NR).
John will be selected to play cricket for the first team (S), he will not be selected for the first team (NS).
The team only plays if it does not rain.

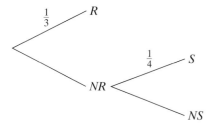

(a) Copy and complete the diagram.

(b) Work out the probability of John playing cricket tomorrow.

(c) Work out the probability of the game being played, but John not being in the team.

6 Three fair plastic discs have a 1 on one side and a 6 on the other.

(a) Use a tree diagram to show all the possible results when the three discs are thrown.

(b) Work out the probability of getting exactly two sixes.

(c) Work out the probability of getting at least two sixes.

7 A small factory employs 5 men and 15 women making bags in two types of leather, leathers A and B. Out of every 20 bags, 14 are made of leather A.
Two different types of bag are made, a shoulder bag and a handbag.
Both types are made in equal numbers.
The government minister for trade (a woman) is to visit the factory and the manager cannot decide which type of bag and material the minister would like to take away as a reminder of her visit, so he picks a bag at random.

(a) Draw a tree diagram to show the likelihood of the bag being made by a man/woman from type A/B leather and being a shoulder/handbag.

(b) Work out the probability that the minister will get a handbag in type A leather made by a man.

(c) Work out the probability that she will either get a shoulder bag made by a man in type B leather, or a handbag made by a woman in type A leather.

8.11 Conditional probability

When the probability of an event depends on a previous event having happened it is called **conditional probability**.

■ **Conditional probability is the probability of *A* given that *B* has already happened.**

This is written as $P(A \mid B)$.

Example 31

John takes his raincoat to school on some days but not on others. He is more likely to take his raincoat if it is raining or if it looks like rain. The probability of John taking his raincoat if it is raining or it looks like rain is 0.9. The probability of John taking his raincoat if it looks as if it is not going to rain is 0.2. Write these as conditional probabilities.

If an event *A* depends on event *B* then

$$P(A \text{ and } B) = P(A) \times P(B \mid A)$$

P(John takes his raincoat *given* that it rains or looks like rain) = 0.9
P(John takes his raincoat *given* that it does not look like rain) = 0.2

Example 32

A delivery from a sub-contractor contains nine similar components. Four of the components are faulty. A component is chosen at random, and not replaced. A second component is then chosen and both components are checked for faults.

(a) Work out the probability that:
 (i) the second component is accepted given the first is accepted,
 (ii) the second component is accepted given the first is faulty.

(b) Draw a tree diagram to represent this information. Include all the possible outcomes and their probabilities.

(c) What is the probability of:
 (i) two acceptable components being selected?
 (ii) the two components both being either accepted or both being rejected?
 (iii) at least one component being accepted?

(a) **(i)** If the first component is accepted there are 4 acceptable and 4 faulty components left.
   $P(second\ acceptable\ given\ first\ acceptable) = \frac{4}{8} = \frac{1}{2}$
 (ii) $P(second\ acceptable\ given\ first\ is\ faulty) = \frac{5}{8}$

(b) Using *F* for faulty and *A* for acceptable:

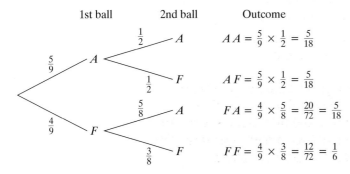

	1st ball	2nd ball	Outcome
		$\frac{1}{2}$ — A	$AA = \frac{5}{9} \times \frac{1}{2} = \frac{5}{18}$
	$\frac{5}{9}$ A		
		$\frac{1}{2}$ — F	$AF = \frac{5}{9} \times \frac{1}{2} = \frac{5}{18}$
		$\frac{5}{8}$ — A	$FA = \frac{4}{9} \times \frac{5}{8} = \frac{20}{72} = \frac{5}{18}$
	$\frac{4}{9}$ F		
		$\frac{3}{8}$ — F	$FF = \frac{4}{9} \times \frac{3}{8} = \frac{12}{72} = \frac{1}{6}$

(c)   **(i)** $P(AA) = \frac{5}{18}$

(ii) $P(AA \text{ or } FF) = P(AA) + P(FF)$

$$= \frac{5}{18} + \frac{1}{6}$$

$$= \frac{8}{18} = \frac{4}{9}$$

(iii) $P(\textit{at least one acceptable component}) = P(AA \text{ or } FA \text{ or } AF)$

$$= 1 - P(FF)$$

$$= 1 - \frac{1}{6}$$

$$= \frac{5}{6}$$

This could also have been found by adding $\frac{5}{18}$, $\frac{5}{18}$ and $\frac{5}{18}$ together.

Exercise 8K

1   A bag contains 5 blue balls and 7 red balls. The balls are taken from the bag one at a time and are not replaced.

(a) Work out the probability that the first ball is blue.

(b) Work out the probability that of the first three balls removed at least two are blue.

2   A box contains 20 marbles of which 5 are red and the rest are white. Marbles are drawn from the bag one at a time without replacement. Work out the probability that:

(a) the second marble drawn is red if the first is not red,

(b) the first three marbles drawn are red,

(c) the first two marbles drawn are red and the third is white.

3   A bowl contains 27 large coloured glass pebbles. Twelve of the pebbles are blue and the rest are red. A customer selects three pebbles at random. The pebbles are not replaced.

(a) Draw a tree diagram.

(b) Work out the probability that the pebbles are all the same colour.

(c) Work out the probability that exactly two pebbles are red.

4   In a group of 30 adults, 7 have blonde hair.

(a) If two adults are chosen at random, work out the probability that they both have blonde hair.

(b) If three adults are chosen at random, work out the probability that exactly two have blonde hair.

5   An investment group of ten financiers have a meeting to vote on their investment strategy for the following three months. Three vote to sell their investments. The rest of the financiers vote not to sell their investments. After the meeting two of them are chosen at random and interviewed on television one after the other.

   (a) Draw a tree diagram to show the ways in which the financiers interviewed could have voted.

   (b) What is the probability that:

      (i) neither voted to sell,

      (ii) one of them voted to sell?

6   A certain medical disease occurs in 2% of the population. A simple screening procedure is available and in 9 cases out of 10 where the patient has the disease it gives positive results. If the patient does not have the disease the test will give a positive result in 5 in 100 cases. Draw a tree diagram and use it to find the probability that a randomly selected person:

   (a) does not have the disease but gives a positive result,

   (b) gives a positive result,

   (c) has the disease.

7   Two computers in a batch of six are known to be faulty but it is not known which two they are. Two machines are picked at random.

   (a) Draw a tree diagram to show all possible results.

   (b) Find the probability that

      (i) they are both faulty,

      (ii) at least one is faulty.

Revision exercise 8

1   A packet contains 12 transistors, three of which are known to be faulty. One transistor is chosen at random and tested.

   (a) What is the probability that the transistor is faulty?

   (b) What is the probability that the transistor is not faulty?

2   The two-way table shows the result of a study by a dentist of 100 patients. Each patient was asked whether or not they smoked and then examined to see how many of their teeth had been taken out.

	Smokers	Non-smokers	Total
One or more teeth taken out	7	13	
No teeth taken out	15		80
Total	22		100

   (a) Copy and complete the table.

   (b) If one of the patients is chosen at random, what is the probability that:

      (i) he/she is a non-smoker,

      (ii) he/she is a smoker who has had no teeth taken out.

   (c) What effect does smoking appear to have on the number of teeth taken out?

3 An insurance company assesses the odds of the houses in a certain village being flooded in any given year at 1000 to 3 against.

   (a) What is the estimated probability of any given house being flooded this year?

   On past experience the average cost to the insurance company of a house being flooded is £6000. The owner of a house in the village wishes to insure his house against flooding for the next year.

   (b) What charge should the insurance company make if its overheads and profits are not included?

4 There are 23 different types of a particular plant (*Ranunculus*) growing in the wild in Britain. 14 of the types have yellow (*Y*) flowers, 14 have 30 or more stamens (*S*), and 13 have both.

   (a) Draw a Venn diagram to show this information.

   (b) What is the probability of a plant, chosen at random from a list of types, has:
   (i) yellow flowers but less than 30 stamens?
   (ii) 30 or more stamens but not yellow flowers?

5 **(a)** What does it mean if two events are said to be 'independent'?

   When a certain make of car breaks down a garage records the breakdown as being due to an electrical fault, due to a mechanical fault or due to other causes. (You can assume that electrical and mechanical faults do not occur together.) The probability of a breakdown in any year due to an electrical fault is 0.1 and to a mechanical fault is 0.05.

   (b) What is the probability in any year of a car of this make breaking down due to an electrical or mechanical fault? What is the probability of the car not breaking down due to one or other of these faults?

   The garage records the rest of the breakdowns, such as running out of petrol, as being due to other causes. The probability of a breakdown due to other causes in any one year is 0.01.

   (c) What is the probability of the car not breaking down in any given year?

6 An aeroplane has two engines. The probability of an engine failing during any flight is 0.01. The plane is able to fly on one engine and the engines fail independently of each other.

   (a) Draw a tree diagram to show the probabilities of the different combinations of engine failures.

   (b) What is the probability of a successful flight?

A second aeroplane has an engine in each wing and a third one in the tail. This aeroplane flies if two engines are working. Assume that the engines have the same reliability as the two-engine plane's engines.

(c) Draw a tree diagram to show the probabilities of the different combinations of engine failure.

(d) What is the probability of the plane making a successful flight?

(e) Would you prefer to travel in the two-engine or three-engine plane?

7 A box contains 50 resistors of which 5 are faulty. A resistor is taken from the box, and not replaced. A second resistor is then taken from the box.

(a) What is the probability that the second resistor is faulty given that the first was?

(b) What is the probability that the second resistor is faulty given that the first was not?

(c) Draw up a tree diagram to show the probability of the different combinations of faulty/non-faulty resistors.

(d) What is the probability of one or less resistors being faulty?

Summary of key points

1 Probability is a numerical measure of the chance of an event happening.
A probability of 0 means the event is impossible.
A probability of 1 means the event is certain to happen.

2 The possible results of a trial are called outcomes.
If outcomes have the same chance of happening they are called equally likely outcomes.

3 Probability of an event $= \dfrac{\text{number of successful outcomes}}{\text{total number of outcomes}}$

4 The estimated probability that an event might happen

$= \dfrac{\text{number of successful outcomes}}{\text{total number of trials}}$

5 In experiments and surveys as the number of trials increases, the nearer your estimate for the probability should be to the true value.

6 Simulation is the imitation of the conditions of a situation by a theoretical study.

7 A list of all possible outcomes is called a sample space.

8 A Venn diagram is a diagram representing a sample space.

9 Outcomes are mutually exclusive if they cannot happen at the same time.

10 For two *mutually exclusive* events A and B

$$P(A \text{ or } B) = P(A) + P(B)$$

11 A set of events is exhaustive if the set contains all possible outcomes.

12 For a set of mutually exclusive, exhaustive events $\Sigma p = 1$.
In particular, $P(\text{not } A) = 1 - P(A)$.

13 Two events are independent if the outcome of one event does not affect the outcome of the other event.

14 For two independent events A and B

$$P(A \text{ and } B) = P(A) \times P(B)$$

15 A tree diagram can be used to aid calculation.

16 Conditional probability is the probability of A given that B has already happened.

9 Probability distributions

9.1 Probability distributions

If you roll a dice there are six possible outcomes and the sample space S = (1, 2, 3, 4, 5, 6). If the dice is unbiased then each outcome is equally likely.

If you let x be a particular outcome:

$$p(x) = \frac{\text{no. of successful outcomes}}{\text{total number of outcomes}} = \frac{1}{6}$$

The different outcomes x and the probability of each outcome would be written:

x:	1	2	3	4	5	6
p(x):	$\frac{1}{6}$	$\frac{1}{6}$	$\frac{1}{6}$	$\frac{1}{6}$	$\frac{1}{6}$	$\frac{1}{6}$

Note:
$\Sigma p = 1$
(The sum of the probabilities is 1.)

The set of all possible outcomes together with their probabilities is called a **probability distribution**.

■ **A probability distribution is a list of all possible outcomes together with their probabilities.**

Example 1

A spinner is designed so that the probability of a 1 occurring is $\frac{1}{2}$. The numbers 2, 3, 4 and 5 occur with frequency k.

(a) Write down the probability distribution of X in terms of k, where X is the number showing when the spinner is spun.

(b) Work out the value of k, and write down the probability distribution of X.

(a)

x:	1	2	3	4	5
p(x):	$\frac{1}{2}$	k	k	k	k

X is the set of values 1, 2, 3, 4, 5.
x is a particular value.

(b)
$$\Sigma p = 1$$
$$\tfrac{1}{2} + k + k + k + k = 1$$
$$4k = 1 - \tfrac{1}{2} = \tfrac{1}{2}$$
$$k = \tfrac{1}{8}$$

The probability distribution is

x:	1	2	3	4	5
p(x):	$\frac{1}{2}$	$\frac{1}{8}$	$\frac{1}{8}$	$\frac{1}{8}$	$\frac{1}{8}$

Exercise 9A

1   Four cars are being filled with petrol at pumps 1 to 4 in a petrol station. X is the pump number of the first car to finish being filled. Shown below is a probability distribution of X.

    x: 1   2   3   4
    $p(x)$: k k k k

Work out the value of k.

2   Clothing is sold in five different size ranges labelled a, b, c, d and e. X represents the size. The probability distribution of X is shown.

    x: a   b   c   d   e
    $p(x)$: 0.1 k 0.3 0.1 0.2

Work out the value of k.

3   There are five faults that can cause a machine to break down. Faults 1 and 5 are likely to happen with probabilities of 0.2 and 0.3 respectively. Faults 2 and 4 are twice as likely to happen as fault 3 is. X is the set of numbers of the faults. If the probability of fault 3 happening is k, draw up the probability distribution of X. Use it to find the value of k.

4   A circular disc is divided into seven sectors. Sector 5 has the same angle as sector 1. Sectors 2 and 6 have twice the area of sector 1. Sectors 3 and 7 have three times the area of sector 1. Sector 4 has four times the area of sector 1. The circular disc is spun, and Y is the set of areas that could come to rest opposite a pointer. Letting s be the probability of $y = 1$, draw up the probability distribution.

(a) Work out the value of s.

(b) What is the probability that $Y < 3$?

9.2 The discrete uniform distribution

Example 2

A six-sided dice is tossed. There are six possible outcomes, the sample space being $X = (1, 2, 3, 4, 5, 6)$. If the dice is unbiased then each outcome is equally likely. Write down the probability distribution of X.

If you let x be a particular outcome:

$$p(x) = \frac{\text{no. of outcomes } x}{\text{total number of outcomes}} = \frac{1}{6} \text{ for all } x$$

The different outcomes x and the probability of each outcome would be written

x: 1   2   3   4   5   6
$p(x)$: $\frac{1}{6}$ $\frac{1}{6}$ $\frac{1}{6}$ $\frac{1}{6}$ $\frac{1}{6}$ $\frac{1}{6}$

A probability distribution such as the one above is known as a **discrete uniform distribution**. It is called discrete because the numbers 1, 2, 3, 4, 5 and 6 are discrete, and uniform because the probabilities are all the same.

■ **A discrete uniform distribution has n distinct outcomes. Each outcome is equally likely. The probability of any given outcome $= \dfrac{1}{n}$.**

Example 3

A taxi is equally likely to pick up 1, 2, 3 or 4 passengers.

(a) Draw up a probability distribution for the number of passengers, A.

(b) What is P(A) < 3?

As the distribution is uniform, all probabilities are the same. $p(a) = \frac{1}{4}$.

(a) a: 1   2   3   4
        $p(a)$: $\frac{1}{4}$ $\frac{1}{4}$ $\frac{1}{4}$ $\frac{1}{4}$

(b) P(a) < 3 = p(1) + p(2) = $\frac{1}{4} + \frac{1}{4} = \frac{1}{2}$

Remember a is a particular outcome.

Exercise 9B

1   Which of the following distributions of X are likely to be modelled by a discrete uniform distribution? Explain your answer.
   (a) X = the heights of a student selected at random from a class of seven-year-olds.
   (b) X = the day of the week on which an adult selected at random was born.
   (c) X = the last digit of a telephone number selected at random from a telephone directory.

2   Given that X is the number showing when a fair dice is thrown, name the distribution of X.

3   A fair octagonal dice has the numbers 1 to 8 on its faces and X is the number that shows when the die is thrown.
   (a) What is the probability distribution of X?
   (b) Work out the probability that $X = 4$.
   (c) Work out the probability that $X < 3$.

4   A dartboard has 20 equal sized sectors numbered 1 to 20. A dart is thrown to land in the number 20 sector. If the dart misses it is thrown again. A discrete uniform distribution is suggested as a model to describe the sector in which the dart lands. Comment on this suggestion.

5   In a draw for a lottery each of the numbers 1 to 50 is equally likely to be picked. X is the number drawn.
   (a) What is the distribution of X?
   (b) What is the probability that $X = 1$?

6   A box of sweets contains equal numbers of orange, red, green, purple and brown sweets. X is the colour of a sweet chosen at random from the box.
   Work out the probability that X = purple.

7   An economist is simulating the daily movement, X points, of a stock exchange indicator. The economist rolls a fair six-sided die and if an odd number is uppermost the indicator is moved down that number of points. If an even number is uppermost he moves it up that number of points.
   (a) Write down the distribution of X.
   (b) What is the probability that $X < -1$ point?

9.3 The binomial distribution

If you throw a dice there are six possible outcomes.
For example, if you want to throw a six then throwing a six is success (s) and any other number counts as a failure (f). If you toss one dice there are two possible results:

s, f

If you toss two dice one after the other **the events are independent** and there are four possible results:

ss, sf, fs, ff

> Remember: two events are independent if the outcome of one event does not affect the outcome of the other.

For three dice there are eight possibilities:

$sss, ssf, sfs, fss, sff, fsf, ffs, fff$

If the order of success and failure is unimportant, terms like ffs, sff and fsf are the same (one six and two other numbers). In which case the above results can be written more concisely as follows:

One dice			s		f	
Two dice		ss		$2sf$		ff
Three dice	sss		$3ssf$		$3sff$	fff

For four and five dice the results would be:

	sss	$4sssf$	$6ssff$	$4sfff$	$ffff$	
$sssss$	$5ssssf$	$10sssff$	$10ssfff$	$5sffff$	$fffff$	

If the probability of success (s) is p and that of failure (f) is q then the corresponding probabilities can be written as:

Since a six is thrown or a six is not thrown, $p + q = 1$.

One dice		p	q		
Two dice		p^2 $2pq$	q^2		
Three dice	p^3 $3p^2q$	$3pq^2$	q^3		
Four dice	p^4 $4p^3q$	$6p^2q^2$	$4pq^3$	q^4	
Five dice	p^5 $5p^4q$	$10p^3q^2$	$10p^2q^3$	$5pq^4$	q^5

The entries in the table are, in the case of

one dice, the terms in the expansion of $(p + q)^1 = p + q$,
of two dice, the terms in the expansion of $(p + q)^2 = p^2 + 2pq + q^2$,
of three dice, the terms in the expansion of $(p + q)^3$
$$= p^3 + 3p^2q + 3pq^2 + q^3, \text{etc.}$$

In an exam you will be given the binomial expression for three or more events, but will be expected to know the expansion for two.

If n dice are thrown the probabilities for each event will be the terms of the expansion of $(p + q)^n$.

These probability distributions are known as **binomial distributions**.

- **A binomial distribution has a fixed number of independent trials n, each of which has only two outcomes (success and failure). The probability of success is p. The probability of failure is q. ($q = 1 - p$)**

This can be written as $B(n, p)$.

- **If n binomial trials are conducted the probability for each event will be terms of the expansion of $(p + q)^n$.**

Example 4

Only 75% of seeds from a particular supplier produce flowers when planted. If four seeds are planted what is the probability of getting (a) three flowers, (b) less than two flowers? You are given that

$$(p + q)^4 = p^4 + 4p^3q + 6p^2q^2 + 4pq^3 + q^4$$

(a) You have four trials so $n = 4$. The probability of success is $p = 75\% = 0.75$ so $q = 1 - 0.75 = 0.25$.

Note the number of successes we are interested in is 3, so you need the term with p^3 in it.

Probability of three flowers $= 4p^3q = 4 \times 0.75^3 \times 0.25 = 0.422$

(b) Probability of less than two flowers $= P(1 \textit{ flower}) + P(0 \textit{ flowers})$
$$= 4pq^3 + q^4$$
$$= 4 \times 0.75 \times 0.25^3 + 0.25^4$$
$$= 0.0508$$

Exercise 9C

1 A distribution X is described as B(12, 0.325).
   (a) What sort of distribution is this?
   (b) What do the numbers 12 and 0.325 represent?

2   A drug cures three people out of every five suffering from a disease.

   (a) Work out the probability of a person given the drug being cured.

   (b) Work out the probability that if four people are given the drug three will be cured.

   You may assume that $(p + q)^4 = p^4 + 4p^3q + 6p^2q^2 + 4pq^3 + q^4$.

3   85% of students who sit a statistics examination pass it. What is the probability that out of a group of three students (a) all three pass, (b) only one passes?

   You may assume that $(p + q)^3 = p^3 + 3p^2q + 3pq^2 + q^3$.

4   Five fair coins are tossed and the total number of heads shown is counted. Work out the probability of:

   (a) one head showing, (b) at least one head showing, (c) the number of heads showing being greater than 4.

   You may assume that $(p + q)^5 = p^5 + 5p^4q + 10p^3q^2 + 10p^2q^3 + 5pq^4 + q^5$.

5   Two people in ten will catch a cold this winter.

   (a) What is the probability that a person will catch a cold this winter?

   (b) In a group of three, what is the probability that at most one person catches a cold this winter.

   You may assume that $(p + q)^3 = p^3 + 3p^2q + 3pq^2 + q^3$.

6   On a certain road, the police stop cars in groups of three taken at random and check the tyres. If one car in ten has faulty tyres, work out the probability that:

   (a) the first three stopped all have faulty tyres,

   (b) none of the first three cars stopped have faulty tyres.

   You may assume that $(p + q)^3 = p^3 + 3p^2q + 3pq^2 + q^3$.

7   The probability of a sheep producing twin lambs is 0.84. If two sheep are selected at random from a flock, what are the probabilities that:

   (a) neither has twins,

   (b) just one has twins?

8   The probability of a water pump being faulty is 0.05. If a sample of five pumps, selected at random, are checked, what are the probabilities that:

   (a) more than three are faulty,

   (b) less than three are faulty,

   You may assume that $(p + q)^5 = p^5 + 5p^4q + 10p^3q^2 + 10p^2q^3 + 5pq^4 + q^5$.

9   The probability that a certain make and model of car breaks down within two years of being bought is 0.1. A hire company buys five of these cars. What is the probability that:

   (a) all five cars break down within the first two years,

   (b) fewer than two of the cars break down within the first two years.

   You may assume that $(p + q)^5 = p^5 + 5p^4q + 10p^3q^2 + 10p^2q^3 + 5pq^4 + q^5$.

10 The probability that a train travelling from Glasgow to Penzance is more than 30 minutes late is 0.15. Three travellers catch trains at different times and dates.
   Work out the probability that:

   (a) all three arrive less than 30 minutes late,

   (b) only one arrives less than 30 minutes late.

   You may assume that $(p + q)^3 = p^3 + 3p^2q + 3pq^2 + q^3$.

9.4 Normal distributions

Suppose you measure the weights of boys: the weight is a **continuous** variable. If you group the weights you can draw a **frequency density histogram**.

> Remember that
> $$\text{frequency density} = \frac{\text{frequency}}{\text{class width}},$$
> and that the area of each bar equals the frequency.

A frequency density histogram for the weights of 100 boys looks like this:

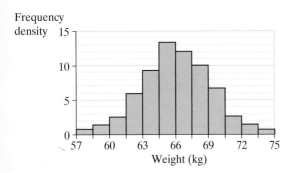

If the number of boys observed was increased to 200 and the class intervals halved, you would get a histogram that looks something like this:

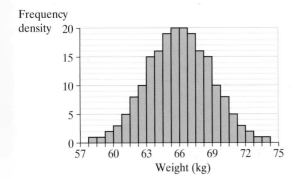

By doubling the number of boys observed and making the class intervals half the size, the outline of the histogram has become smoother. If this process were continued, then the outline of the histogram would eventually have the shape of a smooth curve like this:

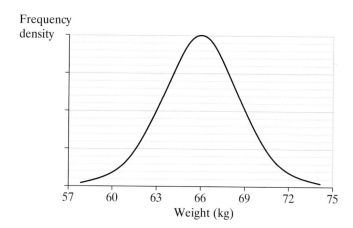

This is shaped rather like a bell and is called a **bell shaped curve**. Because weight is a continuous variable, you require this smooth curve to model boys' weights. It is known as a **continuous probability distribution**.

This distribution is very important in Statistics. Many of the variables that you will meet will have a distribution with this bell shape. If you measured the lengths of 100 cows, the lengths of 100 oak leaves, or the weights of Cox's apples, then a histogram roughly bell shaped would result.

All these are observations of the results of natural processes, and natural processes lead to populations that have this bell shaped curve. Such variables are said to be **normally distributed**. The sketch below shows a **normal distribution** with mean μ.

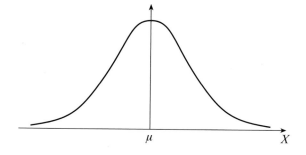

The area under this curve represents the probability of occurrence of the whole distribution. The area = 1.

Two properties of a normal distribution can be summarised as:

■ **The distribution is symmetrical about the mean, μ.**

■ **The mode, median and mean are all equal, due to the symmetry of the distribution.**

9.5 Standard deviation and variance of a normal distribution

A normal distribution will have a mean μ and a standard deviation σ. Different values of μ and σ will give different normal distributions.

Remember: variance $= \sigma^2$

It is helpful to know certain things about the properties of normal distributions. Further properties that you need to know are:

■ **95% of the observations lie within ± two standard deviations of the mean.**

There is an obvious link here with quality control charts. See Example 7.

■ **Virtually all (99.8%) lie within ± three standard deviations of the mean.**

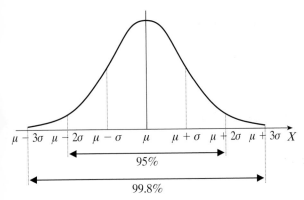

You should note that, because of symmetry, in each case half the area lies on either side of the mean, e.g. 47.5% between the μ and $\mu + 2\sigma$, and 47.5% between μ and $\mu - 2\sigma$. This is shown below.

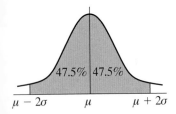 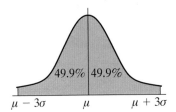

When tackling problems involving normal distributions, it is a good idea to do a sketch showing the area you are trying to find.

Example 5

Sketch on the same axes the normal distributions A and B shown in the table.

	A	B
Mean	15	20
Standard deviation	3	5

When sketching normal distributions you cannot sketch the whole curve since it goes from $-\infty$ to $+\infty$. Instead, sketch three standard deviations either side of the mean.
For A this is $15 \pm (3 \times 3) = 6$ to 24.
For B this is $20 \pm (3 \times 5) = 5$ to 35.

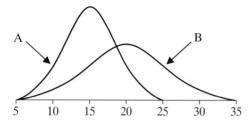

Since the area under each curve has to be 1, the curve with the larger range will have the smaller maximum height.

Example 6

The mark, X out of 100, in an examination is normally distributed. The frequency distribution of X has a mean of 60 marks and a standard deviation of 6 marks. There are 1000 examinees.

(a) **(i)** Work out the mark that is two standard deviations above the mean.
(ii) Between which two values would you expect almost all the marks to lie?

(b) How many marks would you expect to lie between
(i) 48 and 72 marks?
(ii) 60 and 72 marks?

(a) **(i)** $\mu + 2\sigma = 60 + 2 \times 6 = 60 + 12 = 72$
(ii) All marks should lie within $\mu \pm 3\sigma = 60 \pm 3 \times 6 = 42$ to 78.
(b) **(i)** $48 = 60 - 12 = \mu - 2\sigma$ and $72 = 60 + 12 = \mu + 2\sigma$

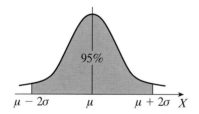

There will be 95% of 1000 = 950 marks between 48 and 72.
(ii) There will be $\frac{1}{2} \times 950 = 475$ between 60 and 72.

Example 7

Samples of size 10 are to be taken from a production line producing jars of coffee. The target weight of the coffee in a jar is 200 g. The means of the samples are normally distributed with a standard deviation of 2 g. Between what limits would you expect:

(a) 95% of the sample means to lie?
(b) 99.8% of the sample means to lie?

(c) What action and warning limits would you use on a control chart for the mean sample weight of the coffee in the jars?

(a) 95% lie between $x = \mu - 2\sigma$ and $x = \mu + 2\sigma$.

$\mu - 2\sigma = 200 - 2 \times 2 = 196$ g and $\mu + 2\sigma = 200 + 2 \times 2 = 204$ g

95% of the means will lie between 196 and 204 g.

(b) 99.8% lie between $x = \mu - 3\sigma$ and $x = \mu + 3\sigma$.

$\mu - 3\sigma = 200 - 3 \times 2 = 194$ g and $\mu + 3\sigma = 200 + 3 \times 2 = 206$ g

99.8% of the means will lie between 194 and 206 g.

(c) Warning limits 196 and 204 g.
Action limits 194 and 206 g.

Example 8

A long life light bulb has a mean life of 12 000 hours and a standard deviation of 300 hours.

(a) Work out the probability that a light bulb chosen at random will:
    (i) last less than 11 400 hours,
    (ii) last between 11 400 and 12 600 hours.

(b) If 5000 light bulbs are tested, estimate the number of light bulbs that would last longer than 12 600 hours.

(a)   **(i)** Number of standard deviations of 11 400 from mean

$$= \frac{11\,400 - 12\,000}{300} = -2$$

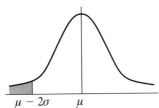

$\mu - 2\sigma$ μ

Probability of failing before $\mu - 2\sigma = 50\% - \dfrac{95\%}{2}$

$$= 50\% - 47.5\%$$
$$= 2.5\% \text{ (or } 0.025)$$

  (ii) Number of standard deviations of 12 600 from mean

$$= \frac{12\,600 - 12\,000}{300} = 2$$

Number of standard deviations of 11 400 from mean

$$= \frac{11\,400 - 12\,000}{300} = -2$$

Probability of lasting between 11 400 and 12 600 hours = 95%.

(b) Probability of a bulb lasting more than 12 600 hours = $50\% - \dfrac{95\%}{2}$

$$= 2.5\%$$

Number of light bulbs lasting more than 12 600 hours from a batch of 5000

$$= 5000 \times 2.5\%$$
$$= 5000 \times \dfrac{2.5}{100}$$
$$= 125 \text{ bulbs}$$

Exercise 9D

1 Which of the following might be modelled by a normal distribution? Explain your answer.

(a) The number of accidents each month on a stretch of road.
(b) The heights of adult females.
(c) The time it takes for a light bulb to burn out.
(d) The distance people travel to work.

2 (a) What relationship exists between the mean, mode and median of a normal distribution?
(b) List the other properties of a normal distribution.

3 The lengths of caterpillars are normally distributed with a mean length of 3.3 cm and a standard deviation of 0.8 cm.
Between which two lengths do nearly all the caterpillars lie?

4 The normal distributions A and B represent the weights of sacks of pre-packed potatoes from two different food producing companies. Sketch on the same axes the normal distributions A and B as described in the table below.

	A	B
Mean (kg)	20	28
Standard deviation (kg)	4	6

5   The random variable X represents the number of sweets in a tube of sweets. X has a distribution with a mean of 18 and a variance of 4.

(a) What is the standard deviation of X?

(b) What value of x is:

(i) two standard deviations below the mean?

(ii) three standard deviations above the mean?

6   The mean speed of vehicles on a road can be modelled by a normal distribution with mean 52.5 km/h and standard deviation 9 km/h. What would be the speed of a vehicle that was travelling at:

(a) two standard deviations above the mean speed?

(b) two standard deviations below the mean speed?

(c) three standard deviations below the mean speed?

(d) three standard deviations above the mean speed?

7   The normal distribution shown below (which has a mean of 52 cm and a standard deviation 6 cm), represents the height of seedling beech hedging plants being sent out from a nursery to customers.

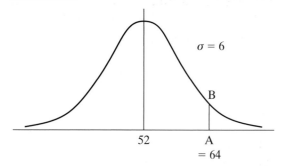

(a) What percentage of the plants will be less than 64 cm high (i.e. lie to the left of line AB)?

(b) What percentage of the plants will be taller than 64 cm (i.e. lie to the right of line AB)?

(c) What percentage of the plants will have heights between 52 and 64 cm?

8   The length X of bamboo canes sold in a garden centre can be modelled by the normal distribution shown in the diagrams below. Work out the probability of a cane chosen at random falling in the shaded areas of each diagram.

(a)

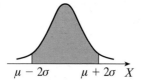

(b)

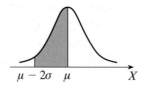

(c) 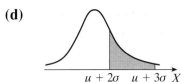 **(d)**

9 The heights of adult men are normally distributed with a mean of μ and standard deviation of σ. What is the probability that a man chosen at random will have a height lying between $\mu + 2\sigma$ and $\mu + 3\sigma$?

10 The mean time it takes factory workers to get to a factory is 35 minutes. The time taken can be modelled by a normal distribution with a standard deviation of 6.5 minutes.
    What percentage of workers will take:
    (a) between 22 and 48 minutes to get to work?
    (b) longer than 48 minutes to get to work?
    (c) If there are 600 factory workers, how many will take between 22 and 48 minutes to get to work?

11 The weights of a group of 1000 school children were found. Their distribution can be modelled by a normal distribution with a mean of 42 kg and a standard deviation of 6 kg.
    (a) What percentage of the students would you expect to find with weights in the range:
        (i) 30 to 54 kg?
        (ii) 24 to 60 kg?
    (b) How many students would you expect to find in each of these size ranges?
    (c) If a student is selected at random what is the probability that the student's weight lies between 24 and 54 kg?

12 It has been found over many years that the temperatures, in °C, for June can be modelled by a normal distribution with a mean of 19 and a standard deviation of 3.5.
    Estimate how many days in June will have a temperature:
    (a) less than 26 °C,
    (b) more than 26 °C,
    (c) between 12 °C and 26 °C.
    Give your answers to the nearest whole number.

13 Tennis balls are to be produced with a mean diameter of 65 mm. Samples are to be taken from the production line and the means of the samples are to have a standard deviation of 0.3 cm. Work out suitable action and warning limits so that 99.8% of the means lie within the action limits, and 95% of the means lie within the warning limits.

14 Television tubes have a mean life of 4000 hours and a standard deviation of 500 hours. Assuming that their life can be modelled by a normal distribution, estimate:

   (a) the probability of a tube lasting less than 3000 hours,

   (b) the probability that a tube will last for between 3000 and 5000 hours.

   (c) In a batch of 10 000 tubes, after how many hours would you expect $2\frac{1}{2}\%$ of the tubes to be still working?

15 The heights of a large number of students can be modelled by a normal distribution with mean 175 cm. 95% of students have heights between 160 and 190 cm. Work out the standard deviation of the students' heights.

16 To test tennis balls they are dropped from a given height and the height they rebound is measured. Balls that rebound less than 128 cm are discarded. Assuming that the rebound height can be modelled by a normal distribution with a mean of 134 cm and a standard deviation of 3 cm, work out how many balls in a batch of 1000 will be rejected.

Revision exercise 9

1 Rashid has in his pocket various numbers of 1p, 2p, 5p and 10p coins. If he puts his hand into his pocket and takes out a coin at random to put in a collection box, then the probability of the coin being a 1p, 2p, 5p or 10p is given by the probability distribution shown:

Coin x:	1p	2p	5p	10p
Probability of x:	$\dfrac{1}{k}$	$\dfrac{5}{k}$	$\dfrac{3}{k}$	$\dfrac{1}{10}$

   (a) Work out the value of k.

   (b) If he has 20 coins in his pocket how many 5p coins are there?

   Emily has equal numbers of each coin in her pocket.

   (c) What name would be given to the distribution of the coins in her pocket?

2 An office is equipped with three new photocopiers. The company that makes the photocopiers states that the probability of any photocopier breaking down in the first three months is 0.01.

   (a) Work out the probability that none of the photocopiers break down in the first three months.

   (b) Work out the probability that two photocopiers break down in the first three months.
   You may use $(p + q)^3 = p^3 + 3p^2q + 3pq^2 + q^3$.

3 Chicken portions produced for a fast food restaurant have weights that are normally distributed with a mean of 160 g and a standard deviation of 10 g.

  (a) What percentage of the portions have weights between:
    (i) 140 and 180 g?   **(ii)** 130 and 190 g?

Portions are packed in boxes of 100 portions.

  (b) How many portions in a box would you expect to weigh between 140 and 190 g?

4 Bags of potatoes for sale in a supermarket have their weights normally distributed with a mean weight of 2 kg and a variance of 144 g. Find the probability that a bag chosen at random will weigh:

  (a) between 1976 g and 2000 g.   **(b)** less than 2024 g.

Summary of key points

1 A probability distribution is a list of all possible outcomes together with their probabilities.

2 A discrete uniform distribution has n distinct outcomes. Each outcome is equally likely.
The probability of any given outcome $= \dfrac{1}{n}$.

3 A binomial distribution has a fixed number of independent trials n, each of which has only two outcomes (success or failure). The probability of success is p.
The probability of failure is q. ($q = 1 - p$)

4 If n binomial trials are conducted the probability for each event will be the terms of the expansion of $(p + q)^n$.

5 A normal curve:
is bell-shaped;
is symmetrical about the mean, μ.
The mode, median and mean are all equal, due to the symmetry of the distribution.

6 For a normal distribution:
95% of the observations lies within $\pm$two standard deviations of the mean.
Virtually all (99.8%) lie within $\pm$three standard deviations of the mean.

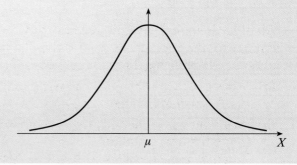

10 Coursework

The coursework for GCSE Statistics may involve you collecting your own data:

or using data from primary and/or secondary sources:

This chapter will give you guidelines on how best to produce a good project.

The coursework flowchart

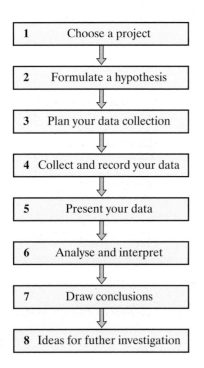

1	Choose a project
2	Formulate a hypothesis
3	Plan your data collection
4	Collect and record your data
5	Present your data
6	Analyse and interpret
7	Draw conclusions
8	Ideas for futher investigation

1. Choose a project

You may be given a project topic and primary/secondary data by your teacher. So you do not need to choose a topic: go to section 2.
When you choose a project topic, choose an idea that interests you. It may be on:

Hobbies or interests

For example:
- the times of goals scored in football matches,
- types of television programmes,
- newspaper comparisons,
- opinion surveys, e.g. health clubs, supermarkets.

Where you study (school, college, etc.)

For example:
- absence rates,
- examination results, e.g. GCSE, GCE, KS3,
- comparisons between students in different years,
- reaction times of boys and girls of different ages.

Local or national data

For example:
- weather,
- traffic,
- unemployment,
- census data.

2. Formulate a hypothesis

When you have decided on your project you must state an idea to be tested. This is called a **hypothesis**.

For example, in a study of the times when goals are scored in football matches, a hypothesis may say:

'A football team is most likely to score a goal just before half time.'
or
'A football team is most likely to concede a goal just after it has scored a goal.'

3. Plan your data collection

Now you have decided on your project topic and what to investigate (your hypothesis), you must collect your data. Decide on the population you are going to use. Decide if you are going to collect primary or secondary data. If you are taking a sample, decide how big it needs to be and choose a suitable sampling method.

Your hypothesis should be chosen so that you can use a variety of statistical techniques from the specification. You should give reasons why you chose each hypothesis.

For the higher marks you need several hypotheses on a related project.

You will investigate to find out whether your hypothesis is right or wrong.

4. Collect and record your data

Now you have decided on your project, what to investigate and your hypothesis, you must collect your data.
You can collect data by:

- experiment, e.g. recording reaction times, observing traffic
- surveys
  – carry out a pilot survey
  – make sure your survey is not biased, do not ask leading questions
  – check your pilot survey is giving you the data you require
  – alter your survey appropriately
  – carry out your survey
- using secondary data, e.g. newspapers, government statistics, searching the internet.

The data collected must be relevant to the hypothesis you wish to investigate. When writing up the project state the method of collection, why you chose it, how well it represents the population and how bias has been avoided.

You now need to record your data.
Use a table that is clear and easy to read.

Sex	Year group	Time (s)
M	10	1.76
F	7	1.34

Other types of table you could use are tally charts and grouped frequency tables.

5. Present your data

You must now present your data using charts or graphs. (A computer package is useful for this stage.) You must select the most appropriate graphs and charts to present your data. For example, to look for relationships use scatter graphs, and use moving averages to look for trends.

If a diagram shows your hypothesis to be obviously incorrect, stop and form a new hypothesis.

When writing up your report, if you have more than one hypothesis you may find it easier to do steps 5 and 6 for the first hypothesis and then do them for the next one.

Always use a sensible degree of accuracy for numerical results. Three significant figures is usually suitable.

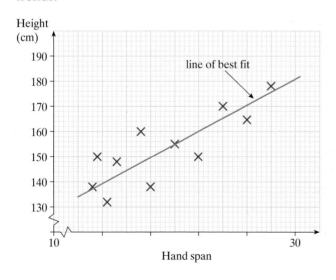

6. Analyse and interpret

To analyse your data you need to select and carry out appropriate calculations of central tendency, dispersion, correlation, time series and index numbers.

You must *interpret* your various tables, charts, diagrams and calculations. Refer back to your hypothesis; do your data support your original thoughts? Write a detailed report. In your report:

- state relationships you found,
- interpret and discuss all diagrams, tables and charts,
- make comparisons of and between data,
- refer back to your hypothesis and support or reject it.

> Sometimes you may find that a hypothesis proves to be false. Do not worry, but try to explain why this could have happened.

7. Draw conclusions

Now you must:

- write out a summary of your findings,
- consider how successful or not you were in achieving your aims,
- discuss any limitations or problems with your project.

8. Ideas for further investigation

Here you should discuss:

- further ideas to extend your project,
- how you would go through the project again, using steps 1–8,
- what new hypothesis you would use and why,
- how you might collect your data in a different way to obtain a better sample.

If you have time, carry out this extension to your project.

Suggestions for projects

Here is a list of project ideas.

- Do students at college have Extra Sensory Perception?
- Investigate the number of sweets in a packet
- For how long do people queue at supermarkets?
- Pulse rates
- Memory tests
- Second-hand car prices
- Employment and crime
- Cinema going, e.g. which age group visits the cinema most often?
- Traffic surveys
- Local community facilities
- Reading book comparisons
- Train statistics
- Weather at holiday resorts
- Spending surveys
- Trees: height, circumference, leaf length
- The national lottery
- Absence from employment
- Fire service call outs

Examination practice papers

Foundation

Section A

1 A student is carrying out a survey of the colours of cars in each car park.
She has three suggestions for methods of recording the data.

A: To write down the colour of each car as she comes to it, e.g. Red, Red, Blue, Green, White, etc.

B: To write down just the first letter of each colour as she comes to it, e.g. R, R, B, G, W, etc.

C: To fill in a tally diagram, e.g.

Colour	Tally
Red	
Blue	
Green	
White	
Silver	
Black	
Other	

(a) Which method should she choose? Give a reason for your answer. (2)

The student also wants to find out about the number of cars entering or leaving the car park. To do this she needs to collect a sample of the cars entering and leaving.

(b) Explain the meaning of the term 'sample'. (1)

She has three suggestions for collecting her sample:
To collect data
  (i) between 10.00 a.m. and 1.00 p.m. on a Saturday,
 (ii) between 10.00 a.m. and 10.30 a.m. on a weekday, or
(iii) at a randomly selected 1 hour period every day.

(c) Which sampling method is best? Give a reason for your answer. (2)

2 The map below shows the reported cases of a disease in six different countries.

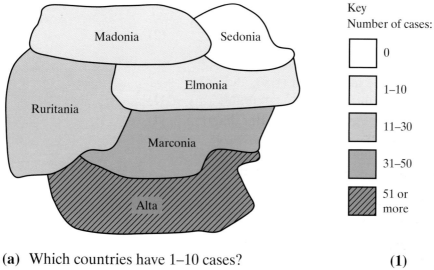

Key
Number of cases:

☐	0
☐	1–10
☐	11–30
☐	31–50
☐	51 or more

(a) Which countries have 1–10 cases? **(1)**

(b) What is a modal class? **(2)**

(c) From which of the 6 countries do you think the disease originated? Give a reason for your answer. **(2)**

3 The bar chart below shows the frequencies with which goals were scored at different times by a football club.

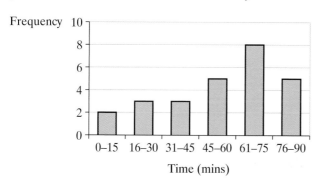

Frequency

Time (mins)

Time (mins)	Frequency
0–15	2
16–30	3
31–45	
46–60	
61–75	
76–90	
Total	

(a) Copy and complete the frequency table on the right. **(2)**

(b) What is the modal class? **(1)**

(c) Work out an estimate of the mean time taken to score a goal. **(3)**

4 A hill walker notes the times it takes him to complete walks of different distances. The results are shown in the table below.

Distance x(km)	30	22	12.5	14	24.5	2.5	5	8	9	17	17.5
Time y (hrs)	9.8	7.0	4.4	4.8	8.8	1.2	2.2	4.0	7.2	6.7	6.1

(a) Plot these points on a scatter diagram. **(2)**

(b) One of these walks was done in very snowy conditions that slowed the walker down. Put a circle around the point on your diagram that refers to this walk. **(1)**

(c) Ignoring the walk done in snowy conditions, work out the mean length of the walks $\bar{x}$. **(2)**

The mean time $\bar{y}$ taken for a walk is 5.5 hours.

(d) On your diagram plot the mean point $(\bar{x}, \bar{y})$, and draw the line of best fit going through the mean point. **(1)**

(e) Describe the correlation seen on your scatter diagram. **(1)**

5   The graph below shows the change in the average price of a house over five years.

Explain two ways in which this graph is misleading. **(2)**

6   The table below shows the nutritional information for a packet of crisps and a packet of dried fruit.

Nutritional information per 100g	Crisps	Dried Fruit
Protein	3.2	2.5
Carbohydrates	55.3	33.9
Fat	31.4	0.4
Fibre	2.2	5.7
Sodium	1.2	0.1
Other	6.7	57.4

Draw composite bar charts to display this data. **(3)**

Foundation

Section B

1 The weights (in grams) of a sample of eggs laid by battery hens are shown below:

56, 57, 63, 66, 68, 68, 69, 70, 71, 73, 74, 74, 75, 76, 77, 78, 79, 81, 83, 85.

The weights (in grams) of a sample of eggs from free-range hens are shown below:

42, 46, 51, 52, 55, 55, 56, 58, 58, 60, 61, 65, 66, 66, 67, 69, 72, 73, 75, 78, 78, 78.

(a) How many eggs were weighed in each sample? **(2)**

(b) Find the median height of the eggs in the sample taken from the battery hens. **(1)**

(c) Draw a back-to-back stem and leaf diagram to represent these data. **(4)**

(d) Which type of hen lays eggs with the widest range of weights? **(1)**

(e) Do free-range hens lay heavier eggs? Justify your answer. **(2)**

2 A road safety officer tests 180 bicycles in a school. The results of a brake efficiency test are given below.

Efficiency	0%–	10%–	20%–	30%–	40%–	50%–	60%–	70%–	80%–
Frequency	2	8	10	23	30	38	32	18	19

(a) Draw a cumulative frequency diagram to display these data. **(4)**

Brakes under 45% efficient are considered unsafe.

(b) Use your diagram to estimate the number of bicycles with unsafe brakes. **(2)**

Brakes under 25% are considered dangerous.

(c) Use your diagram to estimate the number of bicycles that are unsafe but not dangerous **(2)**

3 Copy this probability line:

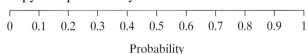

Probability

Mark the probabilities of the following events on your probability line.

A: A person chosen at random is female.

B: A person chosen at random was born in January, February or March.

C: A card chosen at random from a normal pack of cards is not a spade. **(3)**

4 A bag contains five red discs and three yellow discs. A disc is randomly chosen from the bag and then replaced. A second disc is then randomly chosen.

(a) Copy and complete the tree diagram below.

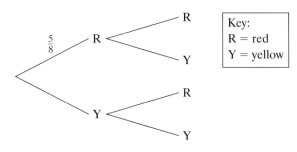

(3)

(b) Find the probability that two yellow discs are drawn. **(2)**

(c) Find the probability that at least one red disc is drawn. **(2)**

5 The lengths (in cm) of 12 goldfish, selected at random from a pet shop, are shown below:

5.4, 6.4, 6.3, 6.3, 7.5, 8.5, 9.0, 9.6, 10.5, 11.3, 14.3, 15.3.

(a) Are the lengths of goldfish continuous or discrete data? **(1)**

(b) Calculate the mean length of goldfish in the sample. **(2)**

The box and whisker diagrams below show the distributions of the lengths of goldfish found in two ponds.

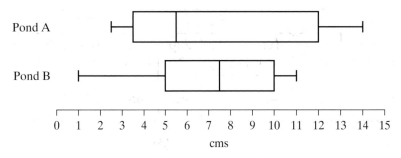

(c) Explain how you can tell from the diagram that the fish in pond B are on average larger than those in pond A. **(1)**

(d) In which pond are the lengths of the goldfish more variable? Give a reason for your answer. **(2)**

(e) In which pond was the smallest fish found? **(1)**

Pond B is larger than Pond A. A biologist claims that goldfish can grow larger in big ponds than they can in smaller ponds.

(f) Do these results support this claim?
Give one reason for your answer. **(2)**

6 The table below gives four categories of household expenditure and the price index for 2003 for 3 of them. (2002 prices = 100).

Category	Frequency
Food	103.1
Fuel and light	98
Personal goods	
Transport	102.6

The prices for personal goods increased by 3.2% between 2002 and 2003.

(a) State the price index for personal goods. **(1)**

(b) Which category was more expensive in 2002 than in 2003? Explain how your know. **(2)**

(c) If the food bill for an average house in 2002 was £2500, estimate the amount the average household spent on food in 2003. **(2)**

An economist wishes to use these data to predict the percentage change in the food bill of an average family between 2003 and 2005.

(d) Explain why his prediction may not be reliable. **(1)**

7 A company works out its sales figures every four months. The results over a three-year period are shown in the table below.

Year	2001			2002			2003		
Period	Jan –Apr	May –Aug	Sep –Dec	Jan –Apr	May –Aug	Sep –Dec	Jan –Apr	May –Aug	Sep –Dec
Sales (in £10 000)	32	48	37	40	52	42	48	56	46

(a) Plot these data as a time series. **(3)**

(b) Calculate three point moving averages for the sales. **(2)**

(c) Plot the moving averages on your graph **(2)**

(d) What is the general trend of the sales over this period of time? Give a reason for your answer. **(2)**

Total marks 80

Examination practice paper

Higher

Section A

1 A fish farmer catches and marks 100 fish from a lake. He returns the fish to the lake. A week later he catches another 100 fish and notes that 25 are marked. He returns these to the lake.

   (a) Estimate the number of fish in the lake. **(3)**

   A year later he captures another 100 fish and notes that only 10 are marked. The farmer uses this as evidence that there are now fewer fish in the lake.

   (b) Give two reasons why he may be wrong. **(2)**

2 School A has 60 teachers and school B has 30 teachers. The table below shows the main teaching subjects of the teachers at each school.

Subject	Number of teachers (School A)	Number of teachers (School B)
English	9	5
Languages	10	4
Mathematics	10	5
Science	12	5
Geography	3	1
History	3	2
PE	5	2
Other	8	6

Karim wants to draw two comparative pie charts to display these data. The pie chart for school B is shown. The radius is 3 cm.

   (a) Work out the radius of the pie chart for school A. **(2)**

   (b) Draw the pie chart for school A. **(3)**

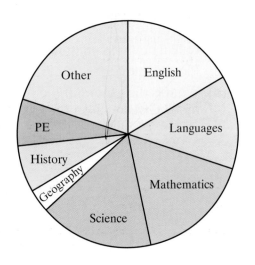

3 The three histograms below represent the distribution of lengths of three types of fish, A, B and C. The histograms are drawn to the same scale.

A

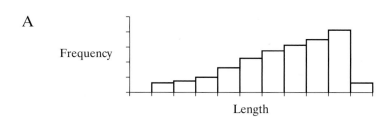

B

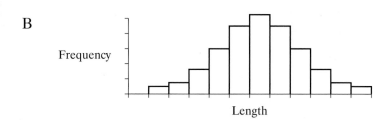

C

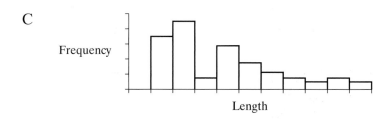

(a) Which type of fish is on average longer?
Give a reason for your answer.

(2)

(b) Choose the correct terms, **normally distributed**,
positively skewed and **negatively skewed** to complete
the sentences below:

   (i) Histogram A is ....

  (ii) Histogram B is ....

 (iii) Histogram C is ....

(3)

(c) For fish B, roughly what percentage of fish have lengths
within two standard deviations of the mean length?

(1)

4 A student wants to carry out a survey of pupils' opinions
about homework.

(a) Give one reason why a pilot survey should be carried out.

(1)

There are 800 pupils in years seven to eleven of a particular school. The table below shows the distribution of the pupils between year groups.

Year	Frequency
7	180
8	180
9	150
10	150
11	140

    (b) If a stratified sample of 100 is to be surveyed, work out how many pupils in each year group needs to be asked. **(4)**

5   **(a)** State which of the following are qualitative and which are quantitative:
    (i) The number of people attending a rock concert.
    (ii) Favourite rock bands. **(2)**

  (b) State which of the following are discrete and which are continuous:
    (i) Weight of apples.
    (ii) The number of pips in apples. **(2)**

6 In a school there are 200 students in Year 11. Of these, 150 own their own mobile phone, 104 own their own computer while 73 own both a computer and a mobile phone.

  (a) Draw a Venn diagram to display these data. **(2)**

  (b) Find the probability that a randomly selected student in Year 11 owns neither a computer nor a mobile phone. **(2)**

7 A fair six-sided dice is thrown together with a biased dice. On the biased dice the probability of a six being thrown is $\frac{1}{3}$. The number of sixes showing on both is noted.

  (a) Copy and complete the following tree diagram.

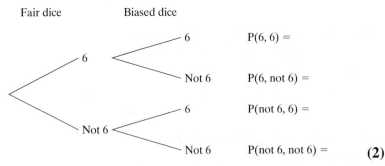

  (b) What is the probability that only one 6 was rolled? **(2)**

  (c) Given that only one six was rolled, what is the probability that it was rolled on the biased dice? **(2)**

Higher

Section B

1   Below is the league table from a local football league.

Team	Points	Goals Scored	Goals conceded
A	48	48	32
B	47	39	26
C	46	53	38
D	45	40	35
E	44	46	34
F	43	52	37
G	42	54	52
H	41	47	33
I	40	38	41
J	35	37	51
K	34	24	23
L	34	30	36
M	31	32	45
N	30	28	31
O	26	27	39
P	24	26	56

(a) Draw a scatter diagram of points against goals conceded.

(3)

The Spearman's rank correlation coefficient of points against goals conceded is −0.405 correct to three significant figures. David says this shows negative correlation.

(b) Is David correct?

(1)

(c) Work out the Spearman's rank correlation coefficient between points and goals scored.

(4)

(d) Comment on your answer to part (c).

(1)

2   The number of road traffic accidents recorded per day at a
    busy junction were as follows:

Accidents per day	Number of days
0	26
1	90
2	57
3	19
4	5
5	3
6	200

(a) Calculate the mean number of accidents per day.   (2)

(b) Calculate the standard deviation of the number of
    accidents per day.   (4)

An earlier study at the same road junction produced the
following results:

Daily road accidents	
Mean	3.2
Standard	1.15

(c) Compare the results of the two studies.   (2)

3   A factory takes samples at regular intervals of the weights of
    bags of potatoes. The samples are plotted on a control chart.
    The chart is shown below.

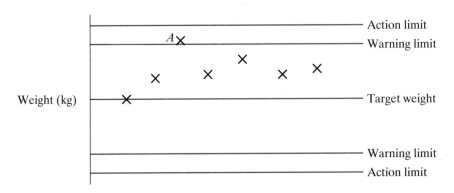

Explain what should have happened at point **A**.   (2)

4 A company works out its sales figures every four months. The results over a three-year period are shown in the table below.

Year	2001			2002			2003		
Period	Jan –Apr	May –Aug	Sep –Dec	Jan –Apr	May –Aug	Sep –Dec	Jan –Apr	May –Aug	Sep –Dec
Sales (in £10 000)	32	48	37	40	52	42	48	56	46

(a) Plot these data on a time series graph. **(3)**

(b) Calculate three point moving averages for the sales and plot them on your graph. **(3)**

(c) Draw a trend line through the moving averages and comment on the trend shown by this data. **(2)**

5 The scatter diagram below shows the value of several Blokeswagen Colf cars plotted against the age of the car.

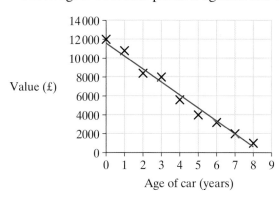

Value (£)

Age of car (years)

(a) Find the equation of the line of best fit, giving your answer in the form $y = ax + b$. **(4)**

(b) Interpret the values of a and b. **(2)**

Shown below are three scatter diagrams, and three equations.

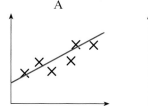
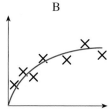
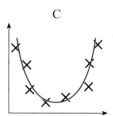

A   B   C

(i) $y = 0.6x^2 + 0.6x + 10$

(ii) $y = 1.6x^2 + 3.3$

(iii) $y = 3.5x^{\frac{1}{2}}$

(c) Write down the most suitable equation for each graph. **(3)**

6 A student records the number of letters delivered to his house each day for 6 months. The number of letters received each day and the frequency with which each occurred is shown.

Number of letters	Frequency
0	21
1	44
2	67
3	29
4	13
5	6
6	2

   (a) What is the median number of letters delivered each day? **(1)**

   (b) Draw a cumulative frequency step polygon for these data. **(4)**

   (c) Find the inter-quartile range for the number of letters delivered. **(2)**

7 The following items are used to calculate the retail price index for the island of Mardsey.
The base time was January 2002.

Item	Weight	Index no. Jan 2003
Food	9	110
Fuel	3	115
Clothes	1	105
Taxes	4	98
Housing	8	110

   (a) Which item is cheaper in 2003 than it was in 2002? **(1)**

   (b) The average household spend £320 on food in January 2002. Use the information in the table to estimate how much they spent on food in January 2003. **(2)**

   (c) Calculate the retail price index for January 2003 using 2002 as a base year. **(4)**

8 A number of walnuts were measured and tasted. They were categorised as sweet or sour. The ordered stem and leaf diagram shows the lengths, in mm, of each type of walnut.

Sour tasting		Sweet tasting
	3	4
	4	7
4	5	5 8
2 4 6 8	6	5 5 9 9
2 3 6 6	7	0 2 4 4 8 8
4 8 8 9	8	0 1
	9	4
	10	0

Key
9 | 4 = 94 mm

(a) Find the median of the sweet tasting walnuts. **(1)**

(b) Find the inter-quartile range of the sweet tasting walnuts. **(2)**

(c) Which of the sweet tasting walnuts are outliers? Explain how you know. **(2)**

The median of the sour tasting walnuts is 73.

(d) Draw box and whisker diagrams for the sour and sweet tasting walnuts on the same scale. **(4)**

(e) Comment on any differences and similarities between the lengths of sweet and sour walnuts. **(2)**

9 The table below shows the percentage of the total area and the percentage of the total population of Wales, England and Scotland.

	Percentage area	Percentage population
England	57	86
Wales	9	5
Scotland	34	9
Total	100	100

(a) Draw two composite percentage bar charts to show these data. **(2)**

(b) Comment on any differences and similarities shown between area and population of the three countries. **(2)**

Total marks 100

Answers

1 (a) (i) discrete
       (ii) quantitative
    (b) colour
2 Continuous, because time is measured.
3 Examples are:
    (a) size   (b) colour   (c) length
4 discrete, whole number
5 Examples are:
    (a) speed
    (b) number of seats being filled
    (c) number of people in it
    (d) enjoyment
6 Appears discrete, because age is usually a whole
    number. Actually continuous because age is not
    exactly (say) 16 years but is 16 years, 3 months, 5
    days, etc.

Exercise 2B

1 By age groups and/or by gender
2 Examples are:
    (a) by age
    (b) by number of bedrooms
    (c) by engine size
    (d) by registration
    (e) by grades
    (f) by male and female
3 e.g. age of car, size of motorcycle engine, colour of
    hair, mpg of cars, etc.
4 Examples are:
    (a) 4 of hearts or similar
    (b) age and weight
    (c) type of car and size of engine
    (d) height and gender
    (e) no. of rooms and size of hotel
    (f) running speeds of various breeds of dogs
5   0–9   0   10–19   9   20–29   8
    30–39 13 40–49   8   50–59   3
    60–69   5 70–79   3   80–89   1

Exercise 2C

1 They are likely to give information they have found.
2 Use published figures about number of sales of
    different types.
    Use published results of questionnaires.
3 Explanatory Height of people
    Response   Time to run distances such as 100 m

4 Individual but these should include
    (i) Primary – new survey on favourite flavours,
        textures, etc.
    (ii) Secondary – use published results.
    (iii) Primary is likely to be more expensive, but more
        reliable and up-to-date.
5 Could
        obtain a postal response,
        pick people at random, but the important
        question is 'Would you go to a new swimming
        pool?'
6 (a) secondary
    (b) James' data is more recent.
7 Individual answers.
8 A personal interview is more likely to produce
    honest answers, but will take longer and be more
    expensive.

Exercise 2D

1 (a) Any reasonable answer, e.g. will reduce wastage
    (b) All the potential users
2 (a) It would destroy the batteries.
    (b) A sample, possibly every 100th battery
3 She could choose women who do not travel, do not
    work, or will lie.
4 (a) Too many, vary all the time
    (b) Could give a view of the whole area

Exercise 2E

1 (a) Any two of: Only numbers 1 to 6 have any
        chance on 1st throw, those not picked first time
        have no future chance, some ($>25 \times 6$) will
        have absolutely no chance …)
    (b) Pick names out of hat.
2 Use integer value of $1200 \times$ RAN#
    40 times, with repeats ignored until the sample = 40.
3 33, 17, 4, 41, 27, 15, 38, 48, 20, 34
4 Individual answers

Exercise 2F

1 (a) Pupils in school
    (b) Only one school used one bus stop.
    (c) Sample names from roll
2 (a) People who live in Britain
    (b) Might be poor or very rich, not typical
    (c) postal

3 Individual answers
4 (a) Convenience
  (b) Biased and unrepresentative
  (c) Own answer, but with a greater emphasis on random
5 (a) Own answer but must say something about views on sport.
  (b) Random sample or similar.
6 (a) 40
  (b) Reception Year: 10, Year 1: 15, Year 2: 15
7 (a) Take every tenth car, observe the aspect you are surveying and record this result. Repeat until you have a sample of 50.
  (b) Merit: easy to carry out, allows time to record results. Disadvantage: not random.
8 Individual answer
9

	Year 7	Year 8	Year 9	Year 10	Year 11
Girls	5.95	6.19	4.76	4.76	3.57
Boys	5.95	5.71	4.76	4.76	3.57
Girls	6	6	5	5	3
Boys	6	6	5	5	3

Exercise 2G

1 (a) Ill defined responses
  (b) How old are you? 0 to 5 ☐ 6 to 10 ☐, etc.
2 (a) Doesn't distinguish between watching a live game or a game on TV, etc.
  (b) Student's answer to include 'How often do you watch a football match?'
   (i) on TV 0, 1, 2, 3, 4 or more times each week
   (ii) visiting a game 0, 1, 2, 3, 4 or more times each week
3 (i) Do you watch TV?   Yes   No
  (ii) How many hours per day? 1 2 3 4 5 6 or more
  (iii) How many days per week? 1 2 3 4 5 6 7
4 (a) Needs a better definition, units of weight and preferably related to age. People may not be honest.
  (b) Ask about age, gender and give boxes (to tick) for weight.
5 Sample from all students at school. Sample size, say 10%. Pilot survey to check the reliability of response and to consider re-wording of questions, etc.
6 (a) Unreliable and too personal
  (b) What is your date of birth? ☐

Exercise 2H

1 (a) Capture–recapture
  (b) Data logging

(c) Matched pairs or control group
  (d) Matched pairs
  (e) Control group
  (f) Matched pairs
2 300
3 Individual answers
4 Before and after
5 Capture, recapture. Not absolutely possible but can use control group to help minimise the effect.

Exercise 2I

1 (a) upper bound £202.50, lower bound £197.50
  (b) upper bound £10 250, lower bound £9750
  (c) upper bound 4.5 m, lower bound 3.5 m
  (d) upper bound 5.5 m, lower bound 4.5 m
  (e) upper bound $125\frac{1}{2}$ miles, lower bound $124\frac{1}{2}$ miles
  (f) upper bound $175\frac{1}{2}$ cm, lower bound $174\frac{1}{2}$ cm
  (g) upper bound 64 years 0 months 0 days, lower bound 63 years 0 months 0 days
  (h) upper bound 202.5 miles, lower bound 197.5 miles
2 (a) upper bound 22.3125 m², lower bound 17.8125 m²
  (b) upper bound 102.5 g, lower bound 97.5 g
  (c) upper bound 274.625 cm³, lower bound 166.375 cm³
  (d) upper bound 15 years 6 months 25 days, lower bound 15 years 6 months 24 days
  (e) upper bound 668 cm³, lower bound 366 cm³ (to the nearest cm³)
3 (a) $4.5 \leqslant x < 5.5$
  (b) $2.5 \leqslant y < 3.5$
  (c) upper bound 1.64, lower bound 1.55
4 (a) $0.68\dot{3}$ (b) $0.340\dot{9}$

Revision exercise 2

1 (a) Qualitative
  (b) Discrete or quantitative
2 (a) All the children in the school
  (b) Stratified sampling
  (c) It suggests the correct answer is 'yes', which would cause bias.
3 Own answers
4 (a) It is about 10% of the population.
  (b) 14 girls, 13 boys
5 (a) No. Class not used at all – not random.
  (b) 14
6 Stratified sample; Year 10: 17 girls, 15 boys; Year 11: 14 girls, 14 boys
7 Own questionnaire
8 (a) 5
  (b) No: 7 Year 7 boys, 3 Year 11 girls (numbers have to be rounded to whole numbers)

Exercise 3A

1

Colour	Tally	Frequency
Black	‖‖ ‖	7
Blue	‖‖ ‖‖	10
Red	‖‖ ‖‖ ‖	12
White	‖‖ ‖‖ ‖‖ ‖‖‖	18
Yellow	‖‖	3
Total		50

2

Number of goals	Tally	Frequency
0	‖‖ ‖	6
1	‖‖ ‖	7
2	‖‖ ‖‖‖	9
3	‖‖‖	4
4	‖‖	3
5	‖	1
Total		30

3 **(a)**

Number of pins	Tally	Frequency
48	‖	1
49	‖‖	3
50	‖‖ ‖‖‖	9
51	‖‖	3
52	‖‖	3
53	‖	1
Total		20

(b) 16 of the boxes have 50 or more, but there are 4 boxes with less than 50, so their claim is false.

4

Word length	Tally	Frequency
1		0
2	‖‖ ‖‖ ‖	11
3	‖‖ ‖‖‖	8
4	‖‖ ‖‖ ‖	12
5	‖‖ ‖‖	10
6	‖	2
7	‖‖	5
8	‖	2
9	‖	1
10	‖	1

5 Depends on the result of the experiment, but table should record number of heads (0, 1, 2, 3) as a tally and a frequency.

Exercise 3B

1 **(a)**

Mark	Tally	Frequency
20–29	‖‖	3
30–39	‖‖ ‖	7
40–49	‖‖ ‖‖ ‖‖‖	13
50–59	‖‖ ‖‖ ‖‖ ‖	16
60–69	‖‖ ‖‖ ‖‖‖	13
70–79	‖‖	5
80–89	‖‖	3
Total		60

(b) 50 students

2 **(a)**

Papers sold	Tally	Frequency
40–44	‖‖ ‖	6
45–49	‖‖	5
50–54	‖‖ ‖	7
55–59	‖‖ ‖‖‖	8
60–64	‖‖‖	4
64–69	‖	1
Total		31

(b) 4 days

3 **(a)**

Age of guest	Tally	Frequency
1–10	‖‖ ‖‖ ‖‖‖	13
11–20	‖‖	3
21–30	‖‖ ‖	6
31–40	‖‖ ‖‖ ‖‖ ‖‖	20
41–50	‖‖	3
51–60	‖	2
61–70	‖‖ ‖‖‖	9
71–80	‖‖‖	4

(b) 31–40 is the most common age group.
(c) She is right: 13 of the children are under 11, but there are only 3 people between 11 and 20.

4 **(a)** These may not be the classes selected by students.

Class (pastime)	Tally	Frequency
Music	||||	4
Sport	✔✔ |	11
Gardening	✔✔ |	11
Reading and writing	✔ |	6
Relaxing	✔	5
Holidays	✔	5
DIY	✔	5
Film and theatre	|||	3
Total		50

(b) Student's own reasons. e.g. You can relax on a holiday, so holidays could join two groups.

5 **(a)**

Mark	Tally	Frequency
1–10		0
11–20	||	2
21–30	|||	3
31–40	✔	5
41–50	✔ |	6
51–60	✔ ||	7
61–70	|	1
71–80	||	2
81–90	|	1
91–100	|||	3
101–110	✔ |	6
111–120	||||	4
Total		40

(b)

Mark	Frequency
1–20	2
21–40	8
41–60	13
61–80	3
81–100	4
101–120	10
Total	40

(c)

Mark	Frequency
1–40	10
41–80	16
81–120	14
Total	40

(d) Part **(b)**. I can see a bunch of marks for beginners, and another for experts.

(e) Part **(c)**. The class intervals are too wide, so it looks like there is a uniform distribution. In part **(a)** the intervals are too narrow so that all the frequencies are small.

Exercise 3C

1

Amount	Tally	Frequency
0–£1.00	|||	3
£1.01–£1.50	✔ ✔	10
£1.51–£2.00	✔ ||||	9
£2.01–£3.00	✔ |	6
£3.01–	||	2
Total		30

2 **(a)** 15 **(b)** 59 **(c)** 21–35

(d)

Age	Tally	Frequency
–20	||||	4
21–25	✔ |||	8
26–30	✔ ✔ ✔	15
31–35	✔ ✔ ✔ ||	17
36–45	||||	4
46–	||	2
Total		50

3 **(a)** Most of the data falls into 1 class

(b) When collecting data you do not know what the highest value will be.

(c) That's 9 programmes a day!

(d) The most common numbers of programmes for watching telly are 9–12 and 17–20, but 13–16 is not a common number of programmes to watch.

4 **(a)** Sue has equal class intervals, while John's are varied.

(b) There are only 3 people below 30.

(c) John. Most ages are between 30 and 49 and John has smaller intervals in this range.

(d) He did not know what age the eldest person would be.

(e) Varied class intervals. Last class open.

Exercise 3D

1 **(a)**

	Adults	Children	Total
Right-handed	32	18	50
Left-handed	15	22	37
Total	47	40	87

(b) 50

(c) Yes, over half the children are left-handed, but only $\frac{1}{3}$ of adults are left-handed.

2

	Lemonade	Orange juice	Total
Girls	3	9	12
Boys	10	6	16
Total	13	15	28

3 **(a)**

	'Butter-side down'	'Butter-side up'	Total
Dropped	26	11	37
Thrown	21	24	45
Total	47	35	82

(b) Butter-side down.

4

	Full Membership	Weekends	Total
Adult	59	4	63
Under 14's	12	25	37
Total	71	29	100

59% of over-14s pay for full membership.

5 **(a)**

	Car	Bus	Cycle	Walk	Other	Total
English	15	4	0	1	0	20
Games	3	1	18	7	3	32
Geography	8	4	1	18	1	32
Maths	28	3	1	1	1	34
Science	16	5	7	6	4	38
Total	70	17	27	33	9	156

(b) 38   **(c)** 70   **(d)** 1

6 **(a)**

	A	B	C	D	E
1		Win	Lose	Draw	Total
2	Team Playing in Blue	367	185	229	781
3	Team Playing in Red	442	229	255	926
4	Total	809	414	484	1707

(b) Cell B4 '=SUM(B2:B3)'
Cell C4 '=SUM(C2:C3)'
Cell D4 '=SUM(D2:D3)'
Cell E4 '=SUM(E2:E3)'
Cell E2 '=SUM(B2:D2)'
Cell E3 '=SUM(B2:B3)'

7 **(a)**

	A	B	C	D	E	F
1		Week one	Week two	Week three	Week four	Total
2	Monday	367	185	410	229	1191
3	Tuesday	442	229	546	255	1472
4	Wednesday	552	338	670	310	1870
5	Thursday	387	286	412	272	1357
6	Friday	298	279	359	183	1119
7	Saturday	646	550	812	435	2443
8	Total	2692	1867	3209	1684	9452

(b) Cell B8 '=SUM(B2:B7)'
Cell C8 '=SUM(C2:C7)'
Cell D8 '=SUM(D2:D7)'
Cell E8 '=SUM(E2:E7)'
Cell F8 '=SUM(F2:F7)'
Cell F2 '=SUM(B2:E2)'
Cell F3 '=SUM(B3:E3)'
Cell F4 '=SUM(B4:E4)'
Cell F5 '=SUM(B5:E5)'
Cell F6 '=SUM(B6:E6)'
Cell F7 '=SUM(B7:E7)'
(c) £2443
(d) £1867
(e) £1930

Exercise 3E

1 **(a)** **(i)** 475 km
      (ii) 174 km
      (iii) 245 km
   (b) 125 + 470 = 595 km
   (c) 245 + 323 + 277 + 470 = 1315 km
2 **(a)** 0730
   (b) 0845
   (c) 1119
   (d) 2040
   (e) **(i)** 12 mins
      (ii) 16 mins
      (iii) 11 mins
      (iv) 50 mins
   (f) 1354
3 **(a)** **(i)** £317
      (ii) £446
      (iii) £266
   (b) 51 +
   (c) 300 − 238 = £62
   (d) 17–25 year old male from area E.
   (e) 36–50 year old female from area B.
4 **(a)** Portendales & The Town
   (b) Hillstone
   (c) Portendales & The Marion
   (d) Hillstone, The Marion
   (e) Portendales
5 Will depend on individual student

Exercise 3F

1 **(a)** Theme park
   (b) 250
   (c) 425
   (d) 162 approx.
   (e) 425 + 250 + 162 + 388 = 1225 pupils approx.

2

A pictogram to show hometowns of supermarket shoppers

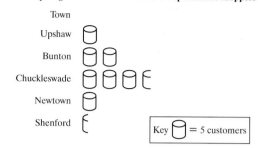

Key = 5 customers

3

A pictogram to show the areas of public spending

Health	
Education	
Transport	
Emergency Services	

Key ☐ = £5 000 000

4 (a) There is no key.
  (b) His fish are drawn bigger than the others.
  (c) Mr Cod – they have 3 fish each. The end of Mr Salmon's fish is the same as the end of Mr Carp's fish, but Mr Carp has more.

Exercise 3G

1 (a) 4
  (b) 7
  (c) 3
  (d) 6 and 9 OR 3 and 11
  (e) $1 + 3 + 4 + 5 + 7 + 6 + 5 + 0 + 1 = 32$
2 (a) 7
  (b) 9
  (c) $7 + 6 + 2 + 3 + 2 + 5 + 3 + 1 + 1 + 1 = 31$
  (d) 0 pets
  (e) Yes, it is very easy to read off the frequencies.
3 (a) A pictogram to show the number of children in 30 families.

Number of children

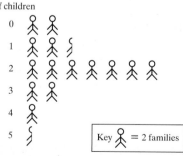

Key 🧍 = 2 families

(b) A bar chart to show the number of children in 30 families.

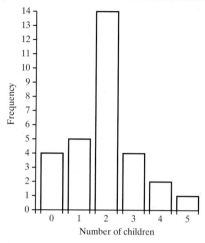

  (c) Opinion
  (d) Bar chart. The exact frequencies can be read from the vertical axis.
4 (a) A bar chart to show hometowns of supermarket shoppers.

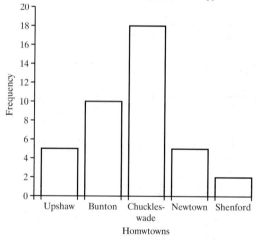

  (b) Chuckleswade.
  (c) This will depend on student's opinion.
5 (a) A pictogram to show lengths of words

Number of letters

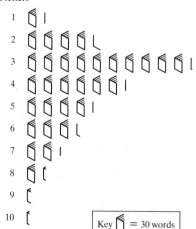

Key 📖 = 30 words

(b) Opinion. (No, the figures need a more accurate method, or Yes, you get a good impression of the common lengths of words).

(c) Line graph → easier to fit on.

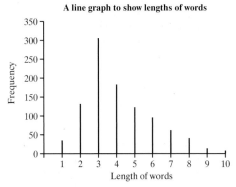

A line graph to show lengths of words

7 (a) It does not start at zero on the vertical axis. It starts on 40, exaggerating the difference between the teams.

(b) The City bar is wider, and coloured darker.

(c) Rovers and City are not side by side, so you cannot tell which bar is higher.

(d) He has not labelled any axes or bars.

Exercise 3H

1 (a) Blue
  (b) Black
  (c) 2 + 3 + 5 = 10
  (d) 7 + 3 + 2 + 5 + 8 = 25
  (e) Wednesday

2 (a) Oranges   (b) Apples and Grapes
  (c) 24 kg   (d) 12 kg
  (e) 24 kg

3 (a)

A bar chart to show sales of scarves

(b)

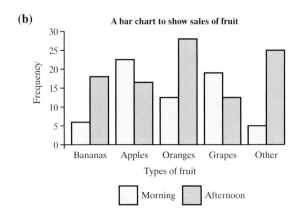

A bar chart to show sales of fruit

Exercise 3I

1 (a)

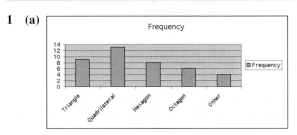

Frequency

2 (b)

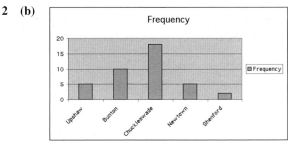

Frequency

(c) Advantage: bar chart is computer generated and looks good.
  Disadvantages: it takes time to set up spreadsheet, and needs access to computer and printer.

3 (b)

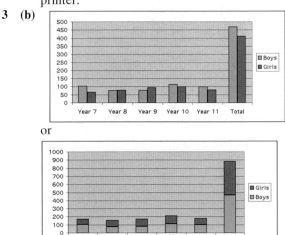

or

Exercise 3J

1 Football = 168° Rugby = 48° Cricket = 84°
   Hockey = 40° Other = 20°

2 (a) $\frac{14}{24} \times 360° = 210°$
   (b) News = 30° Adverts = 60° Music = 210°
   Chat = 45° Weather = 15°

3 (a) $\frac{7}{60}$
   (b) France = 84° Germany = 42° Japan = 78°
   Italy = 48° UK = 108°

4 (a) Duet
   (b) $\frac{3100}{7200} = \frac{31}{72}$
   (c) Duet = 10° Acrobat = 45° Juggler = 90°
   Comedian = 155° Snake charmer = 60°

5 (a) £27
   (b) Food and drink → 42° Clothes → 189°
   CDs → 48° Presents → 81°

6 (a) 5%
   (b) 0 → 54° 1 → 72° 2 → 126° 3 → 90°
   More than 3 → 18°

Exercise 3K

1 Angle = 77° Chart = 85°
   Sector = 71.5°
   Pie = 126.5°

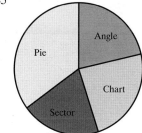

2 (a) 79
   (b) E. Scored.
   (c) I. Missed 77.5°, E. Scored 123°, R. U. Sure 68°,
   P. E. Nalty 55°, Other 35.6°

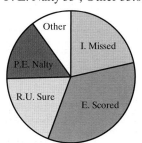

3 (a)

Area of spending	Angle
Health	122°
Education	133°
Transport	32°
Emergency services	73°

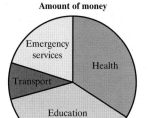

Amount of money

   (b) Student's own answers

4 (a) 28%
   (b)

Number of toilets	Angle
1	169°
2	101°
3	68°
More than 3	22°

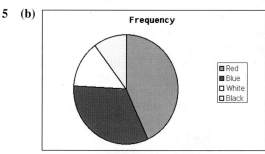

Amount

5 (b)

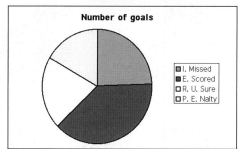

Frequency

6

Number of goals

Exercise 3L

1 (a) 360 − (105 + 125) = 130°
   (b) $\frac{125}{360} \times 72 = 25$ owners
   (c) $\frac{105}{360} \times 72 = 21$ owners

2 (a) Flour = 70 g Butter = 74 g Eggs = 36 g
Sugar = 60 g

(b)

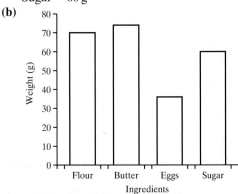

3 (a) 360 ÷ 24 = 15°
(b) 360° − (120 + 30 + 135) = 75°
(c) 120 ÷ 15 = 8 hours
(d) 135 ÷ 15 = 9 hours

4

Colour	Frequency	Angle
Red	8	72°
Blue	14	**126°**
Green	**5**	**45°**
Multi-coloured	**4**	36°
Other	**9**	81°
Total	**40**	**360°**

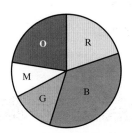

Exercise 3M

1 (a) French
(b) B or C
(c) French
A or A* → 53° B or C → 160°
D or E → 107°
F or G → 40°

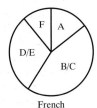
French

Spanish
A or A* → 59° B or C → 213°
D or E → 66° F or G → 22°

Ratio of radii is 1.29 : 1
French : Spanish

Spanish

2 (a)   Urban
(b) Urban
(c) Pinkshire
Agriculture → 34°
Urban → 160°
Woodland → 139°
Water → 27°

Pinkshire

Perkshire
Agriculture → 38°
Urban → 204°
Woodland → 97°
Water → 21°

Perkshire

3 Germany
Beethoven → 38° Mozart → 223°
Handel → 62° Saint-Saëns → 0°
Wagner → 22° Other → 15°

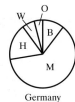
Germany

France
Beethoven → 82° Mozart → 103°
Handel → 61° Saint-Saëns → 92°
Wagner → 12° Other → 10°

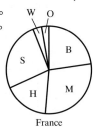
France

4 (a) 320
(b) 63
(c) 27
(d)

Colour	Frequency
Red	63
Blue	45
White	30
Yellow	20
Green	12
Other	10
Total	180

(e) 115
(f)

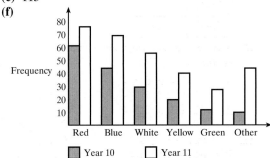

5 (a) A higher proportion of James' class walk, but
there are more pupils in Alex's class overall, and
there are more walkers.
(b) James' class has 25 pupils, and Alex's class has
36 pupils.

Using square roots, make the ratio of the radii of James' pie chart to Alex's 5:6.

James' pie chart

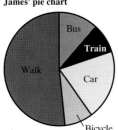

Alex's pie chart

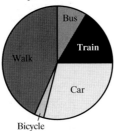

Exercise 3N

1 (a) It is impossible to do 239 sit-ups in this time. It was a lie.
   (b) A stem and leaf diagram to show numbers of sit-ups in a minute

```
0 |
1 | 2 5 6
2 | 3 6 6 7 8 9 9
3 | 2 3 3 3 5 7 8 9
4 | 0 0 1 2 5 8
5 | 3 9
6 | 8
7 | 2 5          Key  3 | 2 = 32 sit-ups
```

   (c) 33

2 (a) 30   (b) 32   (c) 12   (d) 2
   (e)

Age	Frequency
0–4	3
5–9	5
10–14	8
15–19	6
20–24	4
25–29	3
30–34	1
Total	30

   (f) A stem and leaf diagram to show ages of customers

```
0 | 1 2 4 5 6 6 8 9
1 | 1 1 2 2 2 3 4 4 5 5 6 7 8 8
2 | 2 3 3 4 5 5 8
3 | 2          Key  3 | 2 = 32 years old
```

   (g) $0 < \text{age} \leq 10 = 96°$
       $10 < \text{age} \leq 20 = 168°$
       $20 < \text{age} \leq 30 = 84°$
       $30 < \text{age} \leq 40 = 12°$
   (h) Can't see the exact ages any more

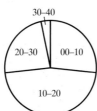

3 (a) **A stem and leaf diagram to show distances travelled.**

```
13 | 6
14 | 2 5
15 | 1 3 4 4 9
16 | 2 2 2 8 9 9
17 | 6 7 8
18 | 0 2
19 | 2 3
20 | 2
```

   Key 13 | 6 = 136
   (b) A bar chart
   (c) The stem and leaf diagram shows all the data. There is no loss of detail.

4 **A stem and leaf diagram to show the number of CDs.**

```
 0 | 0 1 2 4 6 6 7 8
10 | 2 4 4 5 5 8 8 8 8 9
20 | 1 1 2 3 3 5 6 6
30 | 1 1 2
```

   Key 10 | 2 = 12 CDs

5 **A stem and leaf diagram to show the choices of 60 people**

```
10 | 0 2 2 3 3 4 5 6 6 7 9 9
11 | 2 4
12 | 2 3 4 7 8 9
13 | 2 4 4 5
14 | 0 2 3 4 4 5 5 6 7 7 8 8 9 9
15 | 1 1 2 2 3 4 5 5 6 6 7
16 | 2 3
17 | 0 1 1 2 2 3 5 7 8 9
```

   Key 10 | 2 = 102

Revision exercise 3

1 (a) 2
   (b) 23
   (c) 60
   (d) 11 + 23 + 14 + 6 + 3 = 57
   (e) (i) $\frac{90}{360} \times 48 = 12$
       (ii) $\frac{135}{360} \times 48 = 18$
   (f) Squib Street
   (g) No. On Round Street there are more 5 bedroom houses than 4 bedroom houses.
   (h) Crumple Street has the greatest fraction of two bedroom houses.
   (i) The bar chart is easy to read and you can see information at a glance.
       or The pie chart makes it easier to see the proportions of the types of houses.
       or The frequency table gives exact figures.

(j) **(i)**

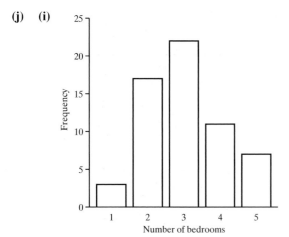

(ii)

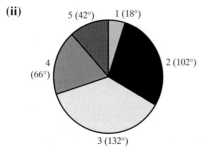

(iii)

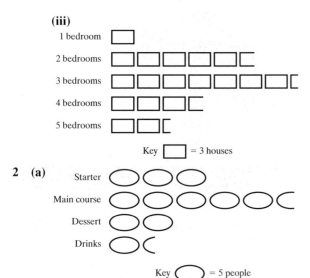

Key ☐ = 3 houses

2 **(a)**

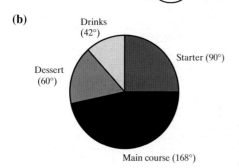

Key ◯ = 5 people

(b)

3 **(a)**

Age	Tally	Frequency
0–9	|||	3
10–19	Ⅷ	5
20–29	Ⅷ Ⅷ ||	12
30–39	Ⅷ Ⅷ		11
40–49	Ⅷ		6
50–59	|||	3
Total		60

(b) The groups would overlap

(c)

```
0 | 3 7 8
1 | 1 2 6 7 7
2 | 1 2 2 4 5 5 6 7 8 8 9 9
3 | 1 3 3 3 5 6 6 6 7 8 8
4 | 1 1 2 4 6 8
5 | 5 6 6
```

Key **2 | 3** = 23 years old.

(d) The exact ages of the customers

4 **(a)**

	Year 7	Year 8	Year 9	Year 10	Year 11	Total
Boys	72	47	71	66	64	320
Girls	63	75	30	55	63	286
Total	135	122	101	121	127	606

(b)

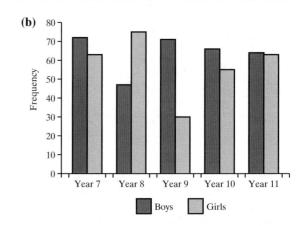

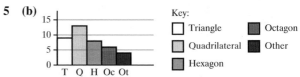

5 **(b)**

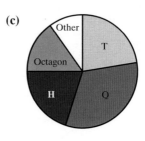

Key:
☐ Triangle   ▨ Octagon
☐ Quadrilateral   ■ Other
▨ Hexagon

(c)

Exercise 4A

1 **(a)** $1 \leq r < 2$

(b)

Rainfall, r (cm)	Tally	Frequency
$0 \leq r < 1$	ⅲ I	6
$1 \leq r < 2$	ⅲ I	6
$2 \leq r < 3$	ⅲ I	6
$3 \leq r < 4$	ⅲ II	7
$4 \leq r < 5$	II	2
$5 \leq r < 6$	IIII	4
Total		31

(c)

2 **(a)** 3.22 kg

(b)

Weight, w (g)	Tally	Frequency
$0 \leq w < 0.5$	II	2
$0.5 \leq w < 1$	III	3
$1 \leq w < 1.5$	ⅲ	5
$1.5 \leq w < 2$	ⅲ IIII	9
$2 \leq w < 2.5$	ⅲ I	6
$2.5 \leq w < 3$	ⅲ I	6
$2 \leq w < 3.5$	III	3

3 **(a)** There is a gap between the groups 1.0 to 1.4 and 1.5 to 1.9, so 1.46 does not fit into any group.

(b) There is an overlap of groups. 1.5 fits into both the groups '1.0 to 1.5' and '1.5 to 2.0'.

(c)

Capacity (litres)
$0 \leq r < 0.5$
$0.5 \leq r < 1$
$1 \leq r < 1.5$
$1.5 \leq r < 2$
$2 \leq r < 2.5$

4

Speed, s (kph)	Tally	Frequency
$30 < s \leq 40$	II	2
$40 < s \leq 50$	ⅲ	5
$50 < s \leq 60$	ⅲ II	7
$60 < s \leq 70$	ⅲ ⅲ	10
$70 < s \leq 80$	ⅲ I	6
Total		30

5 **(a)** The low and high groups can be merged as they have low frequencies.

(b)

Distance, d (metres)	Frequency
$0 < d \leq 20$	7
$20 < d \leq 30$	18
$30 < d \leq 35$	26
$35 < d \leq 40$	37
$40 < d \leq 45$	28
$45 < d \leq 55$	12
$55 < d \leq 70$	2
Total	130

(c)

Exercise 4B

1 **(a)** $47.5 \leq w < 48.5$

(b) All numbers that round to 45 kg do not fit in the same class

(c) $34.5 \leq w < 39.5$, $39.5 \leq w < 44.5$ and so on.

(d)

Weight w (kg)	Tally	Frequency
$34.5 \leq w < 39.5$	II	2
$39.5 \leq w < 44.5$	III	3
$44.5 \leq w < 49.5$	ⅲ I	6
$49.5 \leq w < 54.5$	ⅲ IIII	9
$54.5 \leq w < 59.5$	ⅲ I	6
$59.5 \leq w < 64.5$	III	3
$64.5 \leq w < 69.5$	I	1
Total		30

2 (a)

Time, t (minutes)	Tally	Frequency
$0 \leqslant t \leqslant 9.5$	||	2
$9.5 \leqslant t < 19.5$	⊬	5
$19.5 \leqslant t < 29.5$	⊬ ⊬ ||	12
$29.5 \leqslant t < 39.5$	|||	2
Total		21

(b) Yes $\frac{19}{21}$ is more than 90%

3 (a)

Length, l (cm)	Tally	Frequency
$405 \leqslant l < 455$	|	1
$455 \leqslant l < 505$	||	2
$505 \leqslant l < 555$	⊬ |||	8
$555 \leqslant l < 605$	⊬ ||	7
$605 \leqslant l < 655$	⊬ ⊬	10
$655 \leqslant l < 705$	||||	4
$705 \leqslant l < 755$	||||	4

(b) 2

4 (a) There is a gap between the groups 0 to 9 and 10 to 19, so 9.7 does not fit into any group.

(b) 20.0 seconds to the nearest 0.1 second could have been 19.95 or 20.05 before rounding, and these fall into different groups.

(c) She could have left the last group open. ($t > 50.05$)

Exercise 4C

1

A frequency polygon to show distances travelled

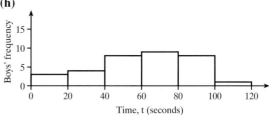

2

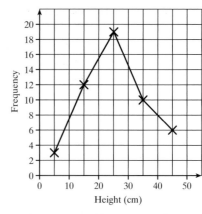

3 (a) 3   **(b)** 7   **(c)** Boy
(d)

Time, t (seconds)	Girls' frequency	Boys' frequency
$0 < t \leqslant 20$	0	3
$20 < t \leqslant 40$	3	4
$40 < t \leqslant 60$	10	8
$60 < t \leqslant 80$	14	9
$80 < t \leqslant 100$	6	8
$100 < t \leqslant 120$	0	1
Total	33	33

(e) Same numbers of each
(f) Girls (more kept going over a minute) *or* both the same (modal times are the same).
(h)

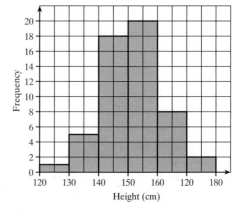

4 (a) A histogram to show heights of Year 7 pupils

(b)

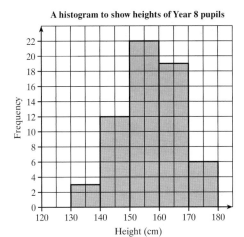

A histogram to show heights of Year 8 pupils

(c)

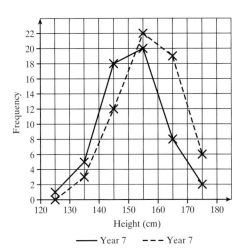

—— Year 7   - - - Year 7

5 (a)

Time, t (mins)	Tally	Frequency
$0.5 \leq t < 10.5$	ⅢⅠ	6
$10.5 \leq t < 20.5$	Ⅲ	5
$20.5 \leq t < 30.5$	ⅢⅠ	6
$30.5 \leq t < 40.5$	ⅢⅠ	4
$40.5 \leq t < 50.5$	ⅢⅠ	6
$50.5 \leq t < 60.5$	Ⅲ	3
Total		30

(b)

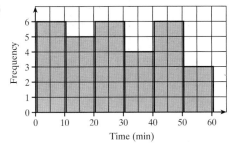

Exercise 4D

1   A histogram to show lengths of songs

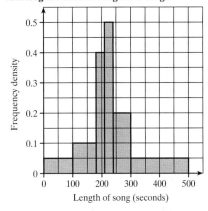

2   A histogram to show ages of guests

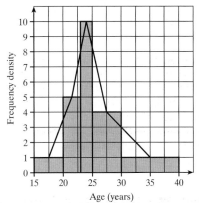

3 (a)

Heights (h cm)	Tally	Frequency
$140.5 \leq h < 150.5$	Ⅲ	5
$150.5 \leq h < 155.5$	Ⅲ Ⅲ	8
$155.5 \leq h < 160.5$	ⅢⅠ	6
$160.5 \leq h < 165.5$	Ⅲ	3
$165.5 \leq h < 170.5$	Ⅲ	3
$170.5 \leq h < 180$	Ⅲ	5
Total		30

(b)

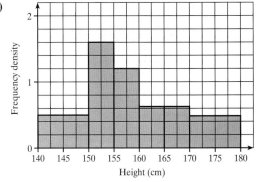

4

Distance travelled, d (cm)	Frequency	Frequency density
$0 < t \leqslant 5$	3	0.6
$5 < t \leqslant 7$	5	2.5
$7 < t \leqslant 8$	4	4
$8 < t \leqslant 9$	6	6
$9 < t \leqslant 10$	3	3
$10 < t \leqslant 15$	6	1.2
$15 < t \leqslant 25$	6	0.6

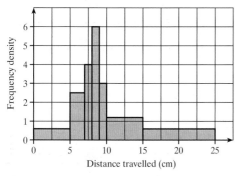

5 Individual survey.

1

Distance, d (metres)	Frequency
$0 \leqslant d < 20$	40
$20 \leqslant d < 35$	75
$35 \leqslant d < 45$	110
$45 \leqslant d < 60$	105
$60 \leqslant d < 65$	10
Total	340

2 (a)

Time, t (seconds)	Frequency
$0 < t \leqslant 5$	56
$5 < t \leqslant 15$	128
$15 < t \leqslant 20$	104
$20 < t \leqslant 30$	160
$30 < t \leqslant 45$	144
$45 < t \leqslant 70$	120
$70 < t \leqslant 80$	24
Total	736

(b) A bar between 70–80, $1\frac{1}{2}$ squares high.

1 (a) 19.9

(b)

```
17 | 3 5 8
18 | 2 3 5 6 8 9
19 | 3 4 6 7 8 9
20 | 2 2 3 5 6 7 8 9
21 | 3 5 8 9
22 | 0 3 8
```
Key 21 | 3 = 21.3 seconds

2 (a) 25

(b) 1.91 m

(c) 17

(d)

Height, h (cm)	Frequency
$1.5 \leqslant h < 1.6$	3
$1.6 \leqslant h < 1.7$	5
$1.7 \leqslant h < 1.8$	9
$1.8 \leqslant h < 1.9$	7
$1.9 \leqslant h < 2.0$	1
Total	25

(e)

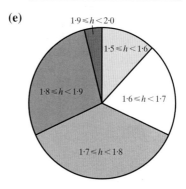

3 (a)

```
0 | 2
1 |
2 | 1 5 2 3 6 3 5 7 8
3 | 4 0 4 3 7 3 1 5 7 7 9 9 7 3
4 | 5 9 7 2 6 3 3 6 8 9
5 | 6 7 9 0 8 0 0 3 3 6 8 9
6 | 0 4 2 3
```
Key 5 | 6 = 5.6 cm

(b)

```
0 | 2
1 |
2 | 1 2 3 3 5 5 6 7 8
3 | 0 1 3 3 3 4 4 5 7 7 7 7 9 9
4 | 2 3 3 5 6 6 7 8 9 9
5 | 0 0 0 3 3 6 6 7 8 8 9 9
6 | 0 2 3 4
```
Key 1 | 2 = 1.2 cm

(c) 3

4 (a) 3.2 seconds
   (b) 21
   (c) $6.3 - 2.0 = 4.3$
   (d) In a frequency table you lose the values, but here you can still see them.

5

0.1	6 7 7 8 8 9 9
0.2	2 2 3 4 4 5 7 7 7 7 7 8 8 8
0.3	2 3 3 5 6 8
0.4	2 4 8 9
0.5	

Key 0.5 | 2 = 0.52

Exercise 4G

1 **A population of pyramid to show percentage of ages**

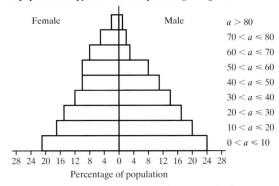

2 (a) B
   (b) Female
   (c) $30 < a \leq 40$
   (d) A
   (e) High mortality rate, high birth rate.

3

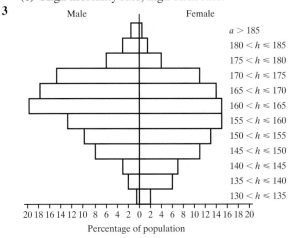

Exercise 4H

1 (a)

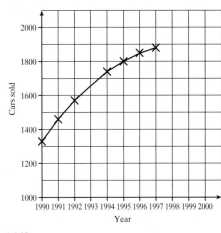

   (b) 1660
   (c) $1998 \approx 1890$
      $1999 \approx 1905$
      $2000 \approx 1920$

2 (a)

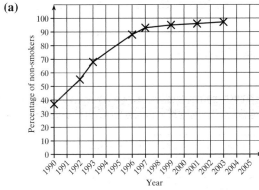

   (b) $1991 \approx 45\%$
      $1994 \approx 76\%$
      $1998 \approx 94\%$
   (c) 97%
   (d) No – it seems to have reached a plateau.
      Perhaps some people will always be smokers. It may never go above 98%.

3 (a)

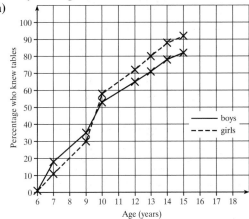

(b) ≈27%

(c) See above

(d) Boys are better when young.

(e) Girls are likely to be better as adults.

(f) Girls = 100%

Boys = 97% (approx)

4 (a)

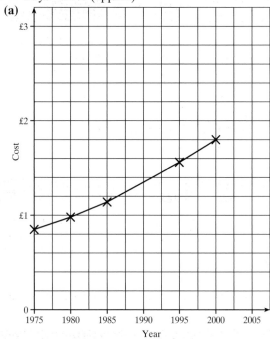

(b) £1.35

(c) £2.10

5 (a)

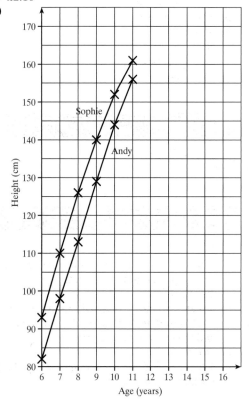

(b) Andy 164 cm   Sophie 167 cm

(c) 13 or 14

Exercise 4I

1

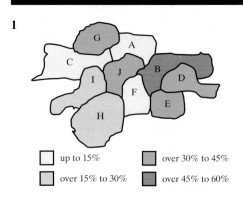

up to 15%   over 30% to 45%

over 15% to 30%   over 45% to 60%

2 (a) A and C

(b) I

(c) B

(d) H, J.

Revision exercise 4

1 (a) This shows that 30 boys balanced the dictionary for between 30 and 40 seconds.

(b)

Time taken, T (seconds)	Frequency
$0 < T \leq 10$	1
$10 < T \leq 20$	6
$20 < T \leq 30$	8
$30 < T \leq 40$	13
$40 < T \leq 50$	2
$50 < T \leq 60$	1
Total	31

(c)

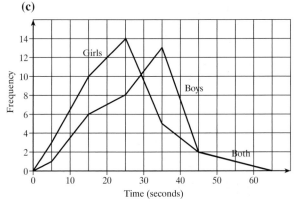

(d) Boys generally balanced for longer (e.g. modal class was higher).

2 (a)

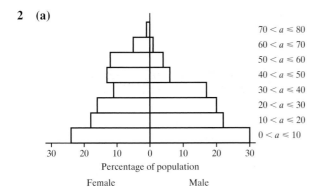

70 < a ≤ 80
60 < a ≤ 70
50 < a ≤ 60
40 < a ≤ 50
30 < a ≤ 40
20 < a ≤ 30
10 < a ≤ 20
0 < a ≤ 10

30   20   10   0   10   20   30

Percentage of population

Female   Male

(b) A True   B Can't tell   C True
     D Can't tell   E Can't tell   F False

3 (a)

Distance (cm)	Frequency
0 < d ≤ 1	5
1 < d ≤ 2	9
2 < d ≤ 3	9
3 < d ≤ 4	7
4 < d ≤ 5	5
5 < d ≤ 6	1
Total	36

(b)

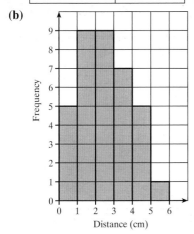

(c) Unordered

```
0 | 5 6 1 0 2
1 | 3 7 3 3 1 9 1 8
2 | 6 2 7 2 1 0 1 8
3 | 7 4 0 3 5 6 2 8 0
4 | 3 2 7 2 5
5 | 2
```

Ordered

```
0 | 0 1 2 5 6
1 | 1 1 3 3 3 7 8 9
2 | 0 1 1 2 2 6 7 8
3 | 0 0 2 3 4 5 6 7 8
4 | 2 2 3 5 7
5 | 2
```

Key 1 | 3 = 1.3 cm

(d) Either 'Histogram because it is easier to see at a glance'
or 'Stem and leaf because you still have the original data.'

Exercise 5A

1 (a) Mode = 9, median = 9, mean = 8.46 (3 sf)
     (b) Mode = 11, median = 9, mean = 8.55 (3 sf)
     (c) Mode = 2, median = 5.5, mean = 6.5
2 Mode = 33, median = 33, mean = 33.42 (2 dp)
3 (a) 520 kg
     (b) 50 kg
4 (a) e.g. 1, 2, 3, 4, 5, 6, 7, 8, 9, 10
     (b) 5, 5, 5, 5, 6, 7, 8, 8, 8, 8

Exercise 5B

1 (a) 13   **(b)** 13   **(c)** 13.26 (2 dp)
2 (a) 3   **(b)** 4   **(c)** 4.29 (3 sf)
3 (a) 20   **(b)** 20   **(c)** 20.48 (2 dp)
4 (a) (i) B
        (ii) C
     (b) Need to give ratings as numbers, not in letters.

Exercise 5C

1 (a) 60 < speed ≤ 70
     (b) 57 mph (approx.)
     (c) 54.9 mph
2 (a) 20 ≤ age < 30
     (b) 31 years (approx.)
     (c) 31.75 years

Exercise 5D

1 3004.9 (1 dp)
2 2.15375

Exercise 5E

1 (a) 14
     (b) Median (or mode) but not mean because 100 is very large and others are not.
2 (a) 20.8 years
     (b) mode = 18, not central within the range
3 (a) Explanation to include mode, median and mean and also differences in price.
4 (a) Ordinary
     (b) The mode is the only average that can be found for non-numerical data.

5 (a) The mode is the only average that can be found for non-numerical data.
   (b) Colour not numerical
6 (a) 25.57 °C (2 dp)
   (b) All numbers are different
   (c) Yes, it's fairly close to all the observed values.

Exercise 5F

1 (a) Jimmy 54.65, Sumreen 58.25
   (b) Sumreen's overall mark is greater than Jimmy's.
2 54.5

Exercise 5G

1 (a) 4.28 (3 sf)
   (b) 1.259 (4 sf)
2 (a) 0
   (b) Because the list of numbers contains 0 and is not typical of the other values.
3 1.233

Exercise 5H

1 91.67 (4 sf)
2 (a) 1.093, 1.098, 1.133, 1.078, 1.218 (4 sf)
   (b) 0.8, 0.802, 0.805, 0.806, 0.82 (3 sf)
   (c) Car price has gone down by almost $\frac{2}{3}$ but house prices increased by a factor of almost 1.8 *or* car prices decreased at a fairly constant rate, and house prices increased at a fairly constant rate.

Exercise 5I

1 (a) range = 8, LQ = 3, UQ = 9, IQR = 6
   (b) range = 57, LQ = 21, UQ = 65, IQR = 44
   (c) range = 11, LQ = 6, UQ = 10, IQR = 4
2 (a)

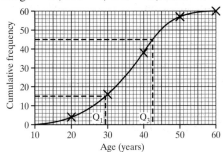

   (b) LQ = 28, UQ = 43, IQR = 15

3 (a)

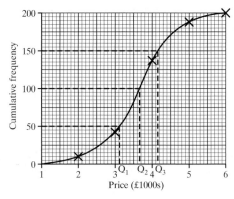

   (b) (i) £3600
      (ii) £4200
      (iii) £3100
      (iv) £1100
   (c) (i) £4500
      (ii) £3400
      (iii) £1100
      (iv) £3500
      (v) £4300
4 (a) £39
   (b)

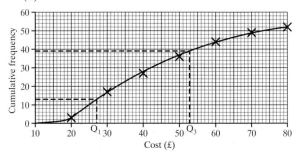

   (c) (i) LQ = £28, UQ = £53
      (ii) 6th decile = £42.50
      (iv) 15th percentile = £23

Exercise 5J

1 (a) (i) 26   (ii) 42
   (b) (i) 40   (ii) 80
   (c)

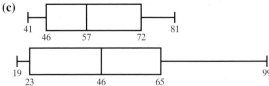

   (d) The girls' marks are more symmetrically distributed. The boys' marks are positively skewed.
2 (a) 75
   (b)

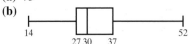

3 (a) median = 38, LQ = 32, UQ = 47
  (b) 76
  (c)

4 (a) LQ = 160, UQ = 180, median = 172
  (b) 120 and 214

Exercise 5K

1 (a)
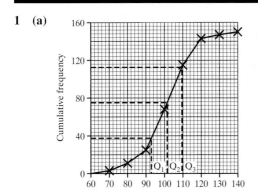

  (b) (i) 102   (ii) 93   (iii) 108   (iv) 15
  (c)
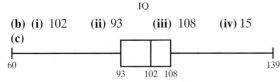

  (d) (i) 104   (ii) 27
2 (a)

  (b) (i) $50 \leqslant s < 60$
    (ii) 54 mph
    (iii) 19 mph
    (iv) 65 mph
     (v) 51 and 58 mph
  (c)
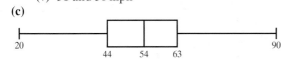

Exercise 5L

1 (a) mean = 8, sd = 2.61 (3 sf)
  (b) mean = $6\frac{2}{3}$, sd = 3.06 (3 sf)
  (c) mean = 4.07 (3 sf), sd = 1.77 (3 sf)
2 (a) 2.61 (3 sf)
  (b) 3.06 (3 sf)   (c) 2.02 (3 sf)

Exercise 5M

1 (a) 41.6
  (b) 110.44
  (c) 10.509
2 (a) mean maximum speed = 99.7 mph
  (b) 9.4
3 (a) 3.48 (3 sf)
  (b) 1.805 (4 sf)   (c) 1.344 (4 sf)
4 (a) mean = 1.65, SD = 4.66 (3 sf)
  (b) mean = 0.1575, SD = 0.596 (3 sf)

Exercise 5N

1 (a) 1.83   (b) 3.25 (3 sf)
2 (a)

	Sean	Theresa	Victoria
History	1.83	−0.5	2.83
Geography	−2	−0.5	1.4

  (b) Victoria is above average in both subjects. Sean is above average in history, but below average in geography. Theresa is below average in both subjects.

Revision exercise 5

1 (a) 32
  (b) 32
  (c) 30.4̇3̇
2 (a) 9
  (b) 9
  (c) 8.48 (3 sf)
3 (a) $20 \leqslant$ age < 30
  (b) (almost) 30 years
  (c) 30.38 years (4 sf)
4 3.042
5 (a) 28.71 °C (4 sf)
  (b) No value occurs more than once.
6 64
7 £11 758.91
8 83.75, 83.58, 80.95, 75

9 (a)

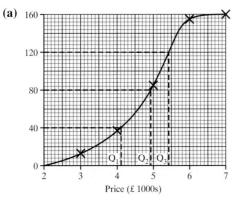

Price (£ 1000s)

(b)

Price (£ 1000s)

(c) (iv) IQR = 5400 − 4100 = 1300;
IQR = 5500 − 4100 = 1400

(d) (i) 4.8   **(ii)** 5.8

(e)

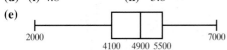

(f) £7350

10 (a) 12   **(b)** 2.45 (3 sf)

11 (a) 1.25, 0.083, 0.5, 1.83 (3 sf)
(b) Jennifer is above average in both subjects
(c) Samuel is above average in both subjects, but well above average in science.

Exercise 6A

1 (a)

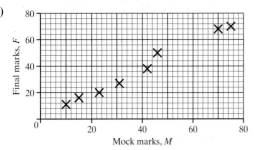

Mock marks, M

(b) Yes

2 (a)

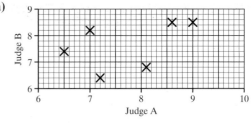

Judge A

(b) There is very slight agreement between the judges.

3 (a)

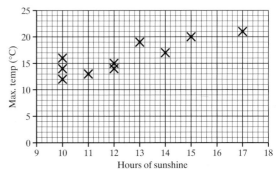

Hours of sunshine

(b) The temperature rises as the number of hours of sunshine increases.

4 (a)

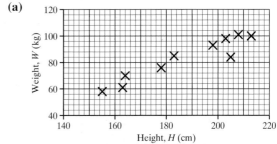

Height, H (cm)

(b) Height and weight are associated. The taller you are the more you weigh.

5 (a)

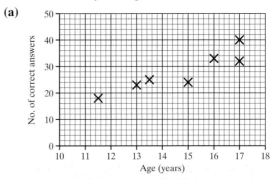

Age (years)

(b) The older a person is the more answers they should get correct.

Exercise 6B

1 (a) No correlation
   (b) Weak positive correlation
   (c) Strong negative correlation
2 (a) Strong positive
   (b) People who do well in the mock examination are likely to do well in the final examination.
3 (a)

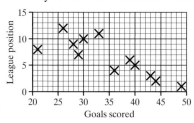

   (b) Weak negative correlation
4 (a)

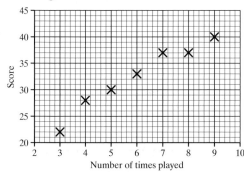

   (b) Strong positive correlation
   (c) The score is likely to increase at the next attempt.
5 (a)

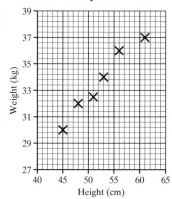

   (b) Strong positive correlation
   (c) The greater the dog's height the more it weighs.

Exercise 6C

1 (a) and (d).
   [In (c), low temperature and snowfall may occur together, but one does not cause the other.]

2 (a) £3500
   (b) 5 years
   (c) Fairly strong negative correlation
   (d) Yes
3 (a) (b)

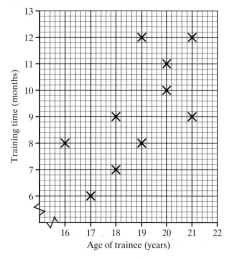

   (c) Evidence of weak positive correlation
   (d) No
4 (a)

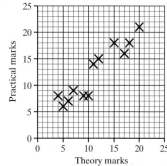

   (b) Fairly strong positive
   (c) No
5 (a)

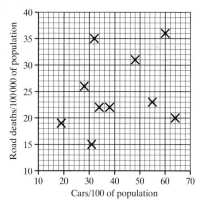

   (b) No correlation
   (c) Yes

Exercise 6D

1 (a)

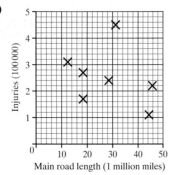

(b) No, there is no correlation.

2 (a) (41, 46)

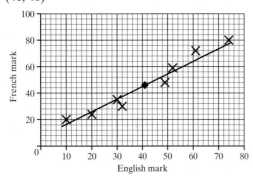

3 (a) (b) and **(d)**

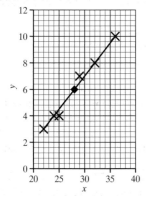

(c) Mean point = (28, 6)

4 (a), (c)

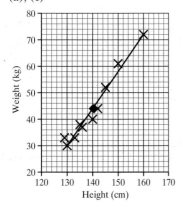

(b) Mean (140, 44)
(d) Strong positive correlation

Exercise 6E

1 (a)

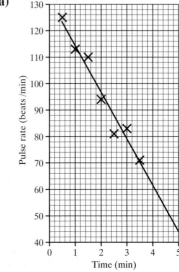

(b) 88 beats/min
(c) 43 beats/min
(d) The first, since interpolation is more accurate than extrapolation.

2 (a), (b)

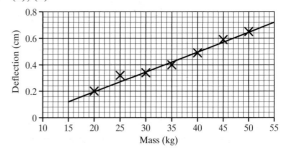

(c) 0.32 cm
(d) 0.14 cm, 0.72 cm
(e) The first, since this is the only one that uses interpolation.

3 (a)

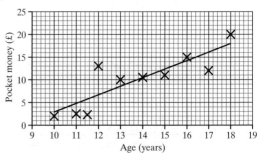

(b) £9.40

(c) People of 25 are usually at work so do not get pocket money.

4 (a) Mean point = (1999, 77.14)

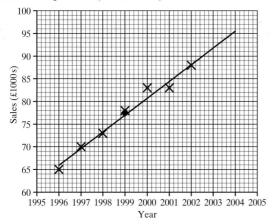

(b) £96 000 (this should be treated with caution since it is an extrapolated value).

5 (a) Mean point (72.5, 102.12)

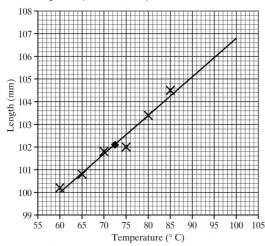

(b) 101.4 mm, 106.8 mm. Interpolation is used for the first so it is reasonably reliable, but the second uses extrapolation so should be treated with caution.

6 (a) Mean point = (1977.5, 16.525)

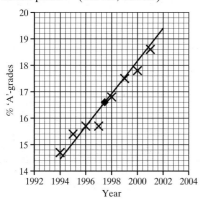

Strong positive correlation

(b) 19.3%

Note: because lines of best fit are drawn by eye, the values for *a* and *b* can be slightly above or below those given. Your answers will also depend on how accurately you can read from the graph.

1 (a) -1.5
  (b) 9.2
  (c) $y = -1.5x + 9.2$
2 (a) $y = 2\frac{2}{3}x - 3\frac{1}{3}$
3 $y = 1.2x + 58.5$
4 (a) Mean point = (13.75, 9.825)

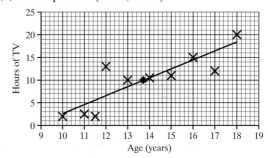

(b) $y = 1.9x + 0.5$
  When $x = 16.5$, $y = 15$
  (c) People of 40 are well outside the range of ages given.
5 (a) 1996 = year 1, then $y = 3.8x + 62.1$
  (b) When $x = 9$, $y = 95.89$ (£95 890)
  Extrapolation is used so treat with caution.
6 (a) $y = 0.17x + 90$
  (i) 101.56 mm   **(ii)** 107 mm.
  The first is more reliable than the second, since first uses interpolation and second uses extrapolation.
  (b) a = the amount the bar expands for each 1 °C rise in temperature, b = the length of the bar at 0 °C.
7 (a) 1994 = year 1 then $y = 1.23x + 13.3$, 19.24%
  (b) You would be extrapolating, and it is unlikely that the percentage would be allowed to reach 50.
8 (a)

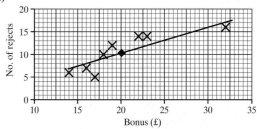

(b) Fairly strong positive correlation – there is evidence to suggest that the manager is correct.
  (c) $y = 0.63x - 2.2$
  (d) £17.8

9 (a) (b)

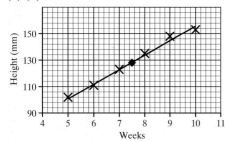

(c) $y = 10.8x + 48$

(d) a = the amount by which the seedling grows in one week, b = the height when the experiment began

(e) 264 mm. Twenty weeks is a long way outside the range of given values so it is unlikely to be very accurate (plants may never grow this high).

Exercise 6G

1 (a) $y = a\sqrt{x} + b$

(b) $y = ax + b$

(c) $y = \dfrac{a}{x} + b$

(d) $y = -ax + b$

2 (a) (b)

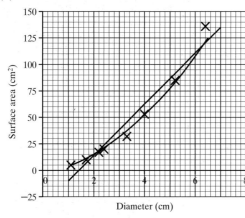

(c) The curve.

3 (a) (b)

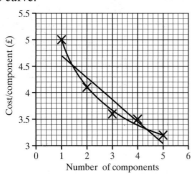

(c) The curve

4 (a)

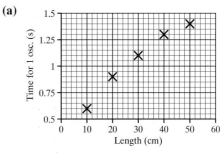

(b) $y = a\sqrt{x} + b$

5 (a)

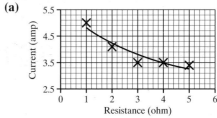

(b) A curve

(c) $y = \dfrac{a}{x} + b$ and $y = ka^x$

Exercise 6H

1

Number	49	44	43	36	40	39	29	28	30	33	26
Rank	1	2	3	6	4	5	9	10	8	7	11

2

Marks after, x	12	22	40	33	18	25	14	4
Rank x	7	4	1	2	5	3	6	8
Marks before, y	10	30	45	12	28	18	19	4
Rank y	7	2	1	6	3	5	4	8

3

Lung damage, x	5	3	7	8	1	4	2	6
Years smoked, y	6	4	7	3	2	8	5	1

4 (a) 0.9371 (4 sf) strong positive correlation

(b) 0.6 medium positive correlation

(c) -0.1905 (4 sf) very weak negative correlation.

5 -0.3714

6 0.7714

7 (a)

Skater	A	B	C	D	E	F
Judge 1	5	2	1	6	3	4
Judge 2	4	3	2	6	1	5

(b) 0.7714

Exercise 6I

1 -0.1152, a very weak negative correlation

2 The two judges are in almost total disagreement.

3 (a) 0.7714
(b) The judges are in moderate agreement.
4 0.4303. There is moderate agreement.
5 (a) 0.881
(b) There is fairly strong agreement.
6 (a) 0.406
(b) There is positive correlation between the ranks, implying that people with long surnames are likely to have long first names. Gemma's belief is false.
7 (a) 0.952
(b) There is strong correlation. The weight after 25 days depends upon the original weight.

Exercise 6J

1 (a)

	A	B	C	D	E	F
Rank 1	3.5	5	2	6	1	3.5
Rank 2	3	5.5	2	5.5	1	4

(b) 0.971
(c) This is strong positive correlation. There is good agreement between the two people.
2 (a) −0.571
(b) This is moderate negative correlation. The more goals scored by a team the fewer the goals scored against the team tend to be.
3 (a)

Student	A	B	C	D	E	F
Rank order	1	2.5	4	2.5	6	5
Rank stats mark	1	2	4.5	3	4.5	6

(b) 0.886
(c) There is fairly strong positive correlation. There is some evidence that the teacher is correct.

Revision exercise 6

1 (a)
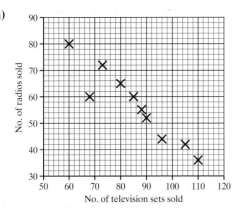
(b) Negative correlation
(c) No. You can have both – ownership of one does not affect ownership of the other.

2 (a) (c)
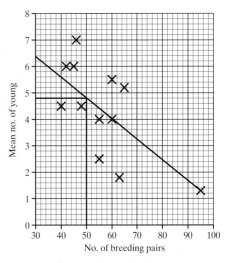
(b) Weak negative
(d) 4.8
3 (a) (b) Mean point (8120, 3665)
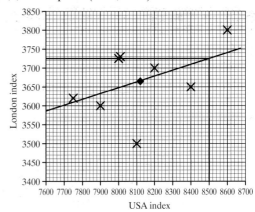
(c) Weak positive
(d) $y = 0.16x + 2375$
(e) 3725
(f) No. This is outside the range and the correlation is weak.
4 (a)
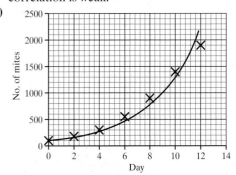
(b) $y = ka^x$ (Law of growth)

5 **(a)**

x	y	Rank x	Rank y	d	d^2
280	90	1	1	0	0
290	95	2	2	0	0
297	110	4	3	1	1
300	125	5	4	1	1
310	140	6	5	1	1
295	145	3	6	−3	9
311	155	7	7	0	0
330	170	9	8	1	1
320	175	8	9	−1	1
				Total	14

(b) $r_S = 0.8833$
(c) Strong positive correlation between the ranks

Exercise 7A

1 **(a)** 3 hours
 (b) Wednesday
 (c) It is the weekend.
2 **(a)**

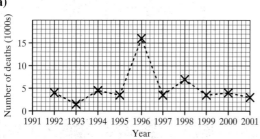

(b) August and December
(c) May
3 **(a)**

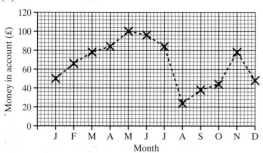

(b) There were many deaths from influenza in 1996.
 There must have been an epidemic.

4 **(a)**

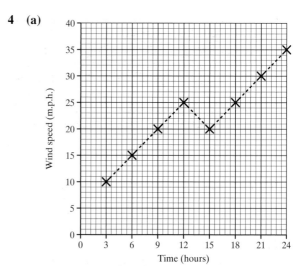

(b) The wind speed rose steadily except for the
 period 12.00 to 15.00 when it dropped slightly.
5 **(a)**

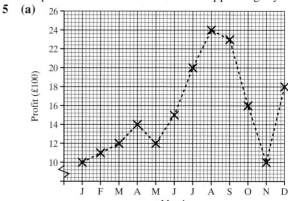

(b) Late summer (August/September)

Exercise 7B

1 **(a)** and **(c)**. **(a)** and **(c)** both alter with time, but **(b)**
 does not.
2 **(a)**

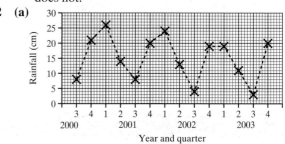

(b) Rainfall seems to be decreasing.

3 **(a)**

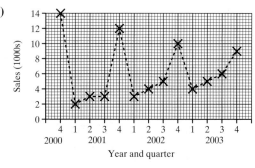

(b) Sales are higher in the fourth quarter, and lowest in the first quarter. Fourth-quarter sales are decreasing from year to year, but first quarter sales are increasing.

4 **(a)**

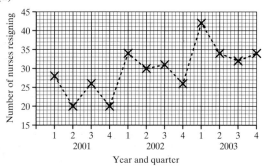

(b) More nurses seem to resign in the first quarter of the year.

(c) Resignations seem to be increasing each year.

5 **(a)**

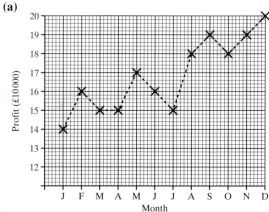

(b) The manager is wrong – profits are rising.

Exercise 7C

1 **(a)** C
   (b) A
   (c) B

2 **(a) (b)**

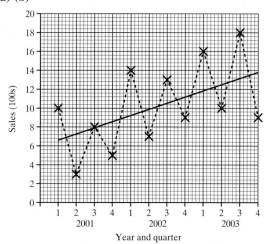

(c) The general trend is for the sales to increase.

3 **(a) (b)**

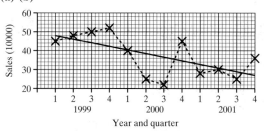

(c) The trend is for sales to decrease.

4 **(a) (b)**

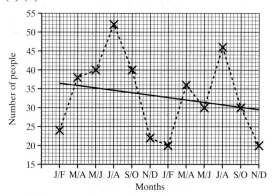

(c) The trend is for the number of people to decrease.

Exercise 7D

1 **(b)** and **(c)**. The number of swimsuits and the number of hours of sunshine rise in the summer and fall in the winter. Bank accounts and the sales of breakfast cereals are not affected by the season of the year.

2

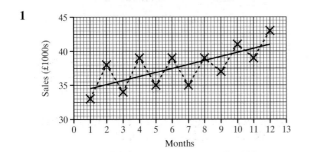

(b) Seasonal variations
(c) The first quarter of the year.

3 (a) 500
(b) 1200
(c) The trend is rising – more people seem to be taking the tour each year.
(d) The tour is most popular in the second quarter of the year and least popular in the fourth quarter of the year.

4 (a) (b)

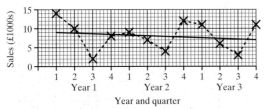

(c) The long-term trend is downwards. Sales are lowest in the third quarter and generally highest in the first quarter.

Exercise 7E

1

Exercise 7F

1 3 point moving averages are 97, 97, 106.3, 111.3, 110.7, 108, 109, 113.3, 120.3, 119, 119.3, 119.7
2 (a) An average taken over successive time periods. A 4 point moving average is given over four successive time periods.

(b)

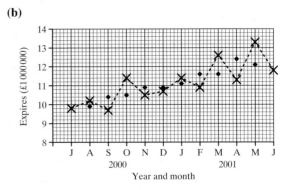

(c) Moving averages 9.9, 10.43, 10.53, 10.87, 10.87, 11, 11.63, 11.6, 12.4, 12.13

3 (a)

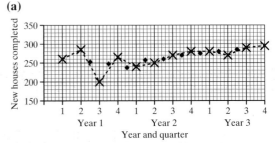

(b) 4 point moving averages 252.5, 247.5, 238.75, 256.25, 260, 270, 275, 280, 283.75
(c) The trend is rising.

4

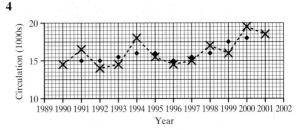

(a) Moving averages 15, 15, 15.5, 16, 16, 15, 15.5, 16, 17.5, 18
(b) The trend is to rise slightly.

5 (a) (b) 20, 20, 22, 24, 24, 25, 26, 27, 30

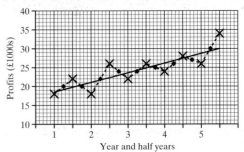

(c) Profits are increasing.
6 (b) Moving averages Firm A 283.3, 350, 413.3, 486.7, 570, 656.7, 726.7
Moving averages Firm B 420, 426.7, 450, 470, 516.7, 546.7, 570

(a) (c) (d)

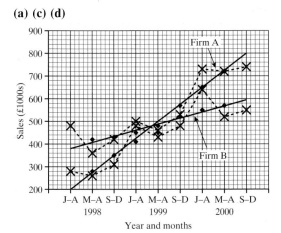

(e) Between April and August 1999

Exercise 7G

1 (a)

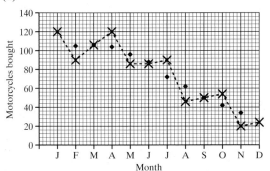

(b) Moving averages 105, 105, 103.3, 96.7, 86.7, 73.3, 61.7, 50, 41.7, 33.3

(c) The trend is falling.

Exercise 7H

Because trend lines are drawn by eye, your readings may vary either side of those given

1 £16.59

2 Totals 3.1, −10.2, 13.1, 7.5; mean seasonal variations 1.55, −5.1, 6.55, 3.75

3 Seasonal variations 4, 14, −4, −4, 1, 9, −4, 0

4 Estimated mean seasonal variations 2.5, 11.5, −4, −2

5 (a) (b) Moving averages 46, 50, 59, 67, 72, 79, 87, 97, 106

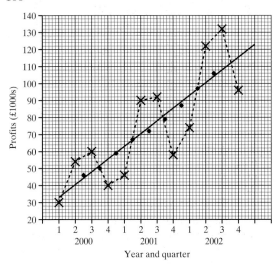

(c) £108 000

6 (a) (c)

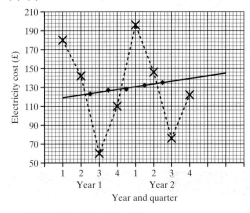

(b) 123, 127, 128, 132, 135

(d) 59, 12

(e) £205, £161

7 (a) The way in which the y variable rises and falls with the seasons, e.g. the sales of ice creams will be at a peak in the summer and low in the winter.

(b) Moving averages 111, 114, 112, 113, 109, 108, 106, 111, 111, 114, 113, 115, 117.

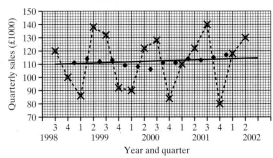

(c) −£11 450, £15 250, £18 160, −£23 150
(d) £133 000, £92 020

8 (a) 664, 681, 694, 712, 734, 756, 780, 799, 822

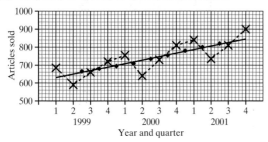

(d)

686	630	56			
590	650		−60		
660	670			−10	
720	690				30
754	710	44			
642	730		−88		
732	750			−18	
808	770				38
842	790	52			
738	810		−72		
808	830			−22	
900	850				50
	Total	152	−220	−50	118
	Mean	$50\frac{2}{3}$	$-73\frac{1}{3}$	$-16\frac{2}{3}$	$39\frac{1}{3}$

(e) 921, 817

Exercise 7I

1 (a) Gradient = £1125/quarter.
(b) The rate at which sales are going up per quarter.

2 (a)

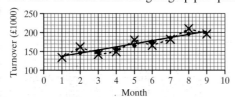

(b) Moving averages 146, 151, 157, 165, 176, 186, 196
(c) $y = 8.54x + 124$
(d) −7.24 × £1000, 17.3 × £1000, −3.16 × £1000
(e) 205 × £1000, 234 × £1000

3 (b) Each period is one third of a year.
(a) (d)

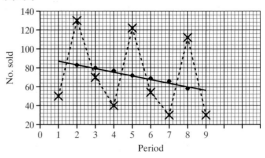

(c) Moving averages 83, 80, 77, 72, 69, 66, 58
(e) Gradient −4. The rate at which sales of size 6 trainers decrease in each 4-monthly period.
(f) −36, 49, −16
(g) 16, 97, 28

Exercise 7J

1 A control chart is used to check that a process's mean, range, variance, etc. stay within set limits, i.e. under control.

2 Components being manufactured may be outside the set limits if either the mean is too large or too small, or if the range is too large.

3

Sample	1	2	3	4	5	6	7	8
Mean	150.4	150	149.8	150.1	150	149.75	149.75	150
Range	0.8	1.7	1.0	1.7	1.4	0.8	1.1	1.4

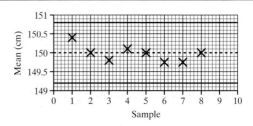

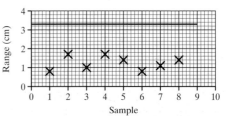

4

Sample	1	2	3	4	5	6	7	8
Mean	49.58	49.98	49.61	49.93	50	49.42	49.9	49.77
Range	1.0	0.29	0.85	0.25	0.08	1.62	0.4	0.62

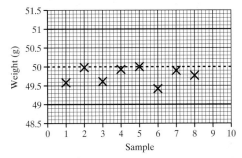

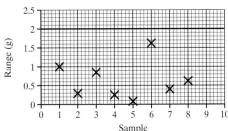

5

Sample	1	2	3	4	5	6	7	8	9	10
Mean	37.78	38.04	37.65	38.04	37.95	37.98	38.26	38.69	39.25	39.79
Range	1.1	0.74	1.12	0.38	1.23	0.81	0.94	0.58	1.03	0.94

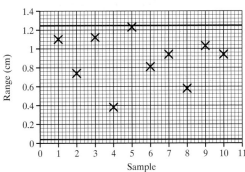

The range is under control but the mean is out of control. The machine should have been stopped after the eighth sample.

Revision exercise 7

1

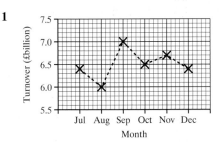

2 (a) (b)

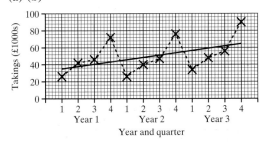

(c) The general trend is for the takings to rise.
(d) Yes. Fourth quarter.
(e) Extra postage before Christmas.

3 (a)

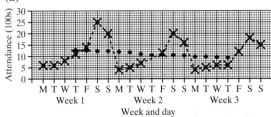

(b) Moving averages 12.86, 12.57, 12.43, 12.29, 12.00, 11.71, 11.00, 10.43, 10.43, 10.43, 10.29, 9.86, 9.86, 9.57, 9.43
(c) The trend is down. The highest attendance is on a Saturday. The lowest attendance is on a Monday.

4 (a)

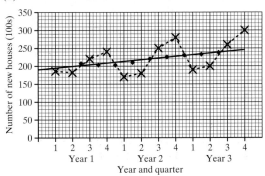

(b) Moving averages 207, 203, 202.5, 210, 220, 225, 230, 232.5, 237.5.
(d) $y = 4.675x + 188$

(e) −29.38, −28.72, 22.61, 47.93
(f) 22 470 houses
5 **(a)** Take another sample.
(b) Stop the process.
(c) Sample means 25.8, 26.1, 25.7, 25.8, 25.9, 25.7, 25.8, 25.7

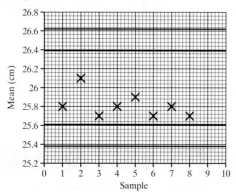

(d) Ranges 1.0, 1.5, 1.4, 0.9, 1.4, 1.5, 1.2, 1.3.

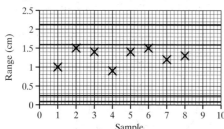

(e) Both mean and range are under control.

Exercise 8A

1 **(a)** certain
(b) evens
(c) unlikely or very unlikely

2

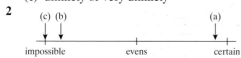

3 **(a)** 0
(b) 1
(c) Probability of W happening is a half (or 0.5 or evens)

4 0.8

5

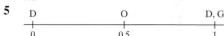

Exercise 8B

1 evens; P(*success*) = P(*failure*) = $\frac{1}{2}$
2 **(a)** 5 : 4 against   **(b)** 2 : 1 against
(c) 7 : 2 against   **(d)** 1 : 2 against

3 **(a)** 5
(b) **(i)** 3 : 2 against
(ii) 3 : 1 against
(iii) 13 : 7 against (7 : 13 on)
4 3 to 2 against
5 $\frac{2}{7}$
6 **(a)** 37 to 3 against
(b) 13 to 27 against
(c) 27 : 13 on

Exercise 8C

1 **(a)** rains; does not rain
(b) 1, 2, 3, 4, 5, 6, 7, 8
(c) point down, point up, on side.
2 **(a)** $\frac{1}{6}$ **(b)** $\frac{4}{6}$ or $\frac{2}{3}$ **(c)** $\frac{2}{6}$ or $\frac{1}{3}$
3 **(a)** $\frac{2}{8}$ or $\frac{1}{4}$ **(b)** $\frac{3}{8}$ **(c)** $\frac{2}{8}$ or $\frac{1}{4}$
(d) $\frac{1}{8}$ **(e)** $\frac{3}{8}$
4 Outcomes that have the same chance of happening.
5 **(a)** $\frac{13}{52}$ or $\frac{1}{4}$ **(b)** $\frac{1}{52}$ **(c)** $\frac{26}{52}$ or $\frac{1}{2}$
(d) $\frac{4}{52}$ or $\frac{1}{13}$
(e) If it is not picked at random then all outcomes are not equally likely.
6 **(a)** $\frac{2}{6}$ or $\frac{1}{3}$ **(b)** $\frac{1}{6}$ **(c)** $\frac{2}{6}$ or $\frac{1}{3}$
7 **(a)** $\frac{25}{50} = \frac{1}{2}$ **(b)** $\frac{16}{50} = \frac{8}{25}$ **(c)** $\frac{16}{50} = \frac{8}{25}$
8 **(a)** Row totals 23, 22, 45; column totals 12, 33, 45
(b) **(i)** $\frac{4}{9}$
(ii) $\frac{2}{9}$
(iii) $\frac{23}{45}$
9 **(a)** Top row 11, 40, 51; middle row 35, 19, 54; bottom row 46, 59, 105
(b) **(i)** $\frac{59}{105}$
(ii) $\frac{8}{21}$
(iii) $\frac{54}{105} = \frac{18}{35}$
10 **(a)** $\frac{3}{8}$
(b) $\frac{5}{8}$

Exercise 8D

1 $\frac{30}{50} = \frac{3}{5}$
2 $\frac{17}{20}$
3 When an experiment or survey is difficult to carry out.

Exercise 8E

1 $\frac{4}{150} = \frac{2}{75}$
2 $\frac{20}{800} = \frac{1}{40}$
3 4
4 £48
5 £100

Exercise 8F

1 (a)

		Coin 1	
		H	**T**
Coin 2	**H**	*HH*	*HT*
	T	*TH*	*TT*

(b) $\frac{2}{4}$ or $\frac{1}{2}$

(c) $\frac{1}{4}$

2 (a)

		Die 1					
		1	**2**	**3**	**4**	**5**	**6**
Die 2	**1**	2	3	4	5	6	7
	2	3	4	5	6	7	8
	3	4	5	6	7	8	9
	4	5	6	7	8	9	10
	5	6	7	8	9	10	11
	6	7	8	9	10	11	12

(b) $\frac{1}{36}$

(c) $\frac{21}{36} = \frac{7}{12}$

(d) $\frac{3}{36} = \frac{1}{12}$

(e) $\frac{18}{36} = \frac{1}{2}$

3 (a)

		Men			
		A	**B**	**C**	**D**
Women	**X**	*XA*	*XB*	*XC*	*XD*
	Y	*YA*	*YB*	*YC*	*YD*

(b) $\frac{2}{8} = \frac{1}{4}$ (c) $\frac{1}{8}$ (d) $\frac{3}{8}$

4 (a)

		London–Reading					
		1	**2**	**3**	**4**	**5**	**6**
Reading–Swindon	**1**	2	3	4	5	6	7
	2	3	4	5	6	7	8
	3	4	5	6	7	8	9
	4	5	6	7	8	9	10

(b) $\frac{3}{24} = \frac{1}{8}$

(c) $\frac{1}{24}$

(d) $\frac{14}{24} = \frac{7}{12}$

5 (a)

	Blue	Red	Green
Blue	BB	BR	BG
Red	RB	RR	RG
Green	GB	GR	GG

(b) $\frac{1}{9}$

(c) $\frac{3}{9} = \frac{1}{3}$

(d) $\frac{1}{9}$

6 (a)

1 1 1						
1 1 2		1 2 1		2 1 1		
1 2 2		2 1 2		2 2 1		
2 2 2						

(b) $\frac{1}{8}$ (c) 0   (d) $\frac{3}{8}$ (e) $\frac{7}{8}$

7 (a)

ABC	ACB
BAC	BCA
CAB	CBA

(b) $\frac{1}{2}$

(c) $\frac{1}{3}$

Exercise 8G

1 (a) $x = 28$; the number of students who study Spanish but not French

(b) $x = 18$; the number of students who study both Geography and History

(c) $x = 3$; the number of students who study neither Maths nor Science.

2 (a)

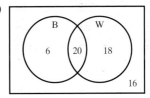

(b) $x = 18$

(c) 26

(d) $\frac{16}{60} = \frac{4}{15}$

3 (a)

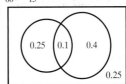

(b)

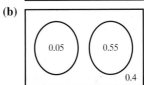

(c)

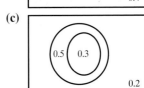

(d)

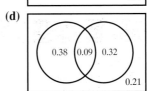

4 (a)

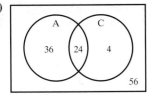

(b) $\frac{56}{120} = \frac{7}{15}$

5 (a)

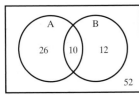

(b) 52%
(c) 26%

6 (a)

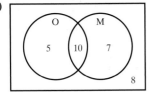

(b) 17
(c) $\frac{7}{30}$

7 (a)

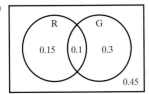

(b) 0.25
(c) 9 parrots

8 (a)

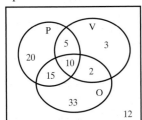

(b) $\frac{3}{25}$
(c) $\frac{20}{100} = 0.2$ or $\frac{1}{5}$

9 $x = 0.2, y = 0.2, z = 0.05$

10 (a)

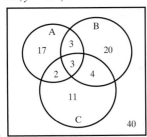

(b) (i) $\frac{3}{5}$ **(ii)** $\frac{12}{25}$ **(iii)** $\frac{2}{5}$

Exercise 8H

1 **(a), (c)**
2 **(a)** 0.6   **(b)** 0.5   **(c)** 0.7
3 **(a), (c)**
4 **(b)**
5 **(a)** $\frac{1}{8}$ **(b)** $\frac{1}{8}$ **(c)** 1
6 0.4
7 **(a)** 15   **(b)** $\frac{10}{30} = \frac{1}{3}$ **(c)** $\frac{5}{30} = \frac{1}{6}$
  (d) $\frac{1}{2}$ **(e)** $\frac{5}{6}$ **(f)** $\frac{2}{3}$
8 **(a)** 0.75   **(b)** 0.95   **(c)** 0.8

Exercise 8I

1 **(a) and (b)**
2 **(a)** 0.06   **(b)** 0.08   **(c)** 0.12   **(d)** 0.58
3 **(a)** Yes   **(b)** 0.42
4 **(a)** $\frac{4}{35}$ **(b)** $\frac{6}{7}$ **(c)** $\frac{6}{35}$
5 **(a)** $\frac{1}{480}$
  (b) $\frac{1}{18}$
6 0.006
7 **(a)** 0.225
  (b) 0.8
  (c) 0.24

Exercise 8J

1 **(a)** $\frac{1}{2}$
  (b) $\frac{1}{2}$
  (c)

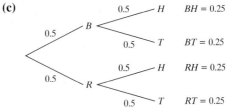

  (d) 0.25
2 **(a)**

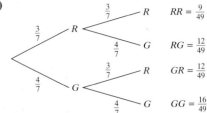

  (b) $\frac{9}{49}$
  (c) $\frac{24}{49}$
3 **(a)** $\frac{6}{8} = \frac{3}{4}$
  (b) $\frac{6}{12} = \frac{1}{2}$

(c)

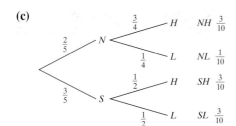

(d) $\frac{3}{10}$

(e) $\frac{2}{5}$

4 (a)

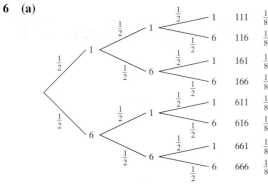

5 (a)

R $\qquad$ R = $\frac{1}{3}$

$\frac{1}{3}$

$\frac{2}{3}$ NR $\qquad$ $\frac{1}{4}$ S $\qquad$ NRS = $\frac{1}{6}$

$\qquad$ $\frac{3}{4}$ NS $\qquad$ NRNS = $\frac{1}{2}$

(b) $\frac{1}{6}$

(c) $\frac{1}{2}$

6 (a)

			111	$\frac{1}{8}$
			116	$\frac{1}{8}$
			161	$\frac{1}{8}$
			166	$\frac{1}{8}$
			611	$\frac{1}{8}$
			616	$\frac{1}{8}$
			661	$\frac{1}{8}$
			666	$\frac{1}{8}$

(b) $\frac{3}{8}$ $\qquad$ **(c)** $\frac{1}{2}$

7 (a)

	S	MAS
	H	MAH
	S	MBS
	H	MBH
	S	WAS
	H	WAH
	S	WBS
	H	WBH

(b) $\frac{7}{80}$ $\qquad$ **(c)** $\frac{3}{10}$

Exercise 8K

1 (a) $\frac{5}{12}$ $\qquad$ **(b)** $\frac{4}{11}$

2 (a) $\frac{5}{19}$ $\qquad$ **(b)** $\frac{1}{114}$ $\qquad$ **(c)** $\frac{5}{114}$

3 (a)

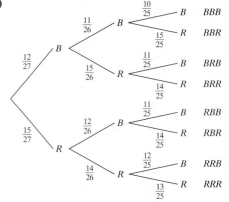

(b) $BBB + RRR = \frac{81}{351} = \frac{3}{13}$

(c) $RRB + RBR + BRR = \frac{28}{65}$

4 (a) $\frac{7}{145}$

(b) $\frac{483}{4060} = \frac{69}{580}$

5 (a)

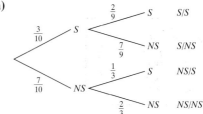

(b) (i) $\frac{14}{30} = \frac{7}{15}$ $\qquad$ **(ii)** $\frac{42}{90} = \frac{7}{15}$

6

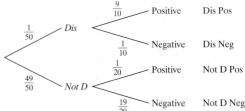

(a) $\frac{49}{1000}$ $\qquad$ **(b)** $\frac{67}{1000}$ $\qquad$ **(c)** $\frac{1}{50}$

7 (a)

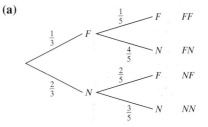

(b) (i) $P(FF) = \frac{2}{30} = \frac{1}{15}$

(ii) $P(FF) + P(NF) + P(FN) = 1 - P(NN)$
$= \frac{18}{30} = \frac{3}{5}$

Revision exercise 8

1 (a) 0.25 or $\frac{1}{4}$ (b) 0.75 or $\frac{3}{4}$

2 (a)

	Smokers	Non-smokers	Total
One or more teeth extracted	7	13	**20**
No teeth extracted	15	**65**	80
Total	22	**78**	100

  (b) (i) $\frac{78}{100} = 0.78$

     (ii) $\frac{15}{100} = 0.15$

  (c) 7 out of 22 smokers (31.8%) have one or more teeth taken out while 13 out of 78 non-smokers (16.6%) have one or more teeth taken out. Smoking is bad for your teeth.

3 (a) $\frac{3}{1003} \approx 0.003$

  (b) £18 or £17.95 if the correct probability is used

4 (a)

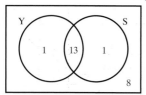

  (b) (i) $\frac{1}{23} = 0.043$ (3 dp)

     (ii) $\frac{1}{23} = 0.043$ (3 dp)

5 (a) The outcome of one event does not affect the outcome of the other event.

  (b) 0.15, 0.85   (c) 0.84

6 (a)

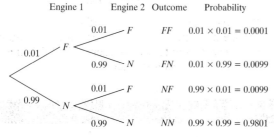

  (b) 0.9999

  (c)

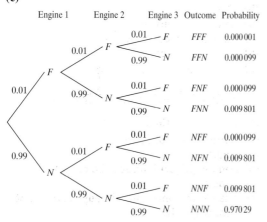

  (d) 0.999 702

  (e) Two engine

7 (a) $\frac{4}{49} = 0.082$ (3 dp)

  (b) $\frac{5}{49} = 0.102$ (3 dp)

  (c)

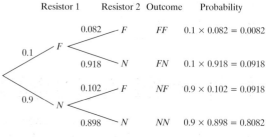

  (d) 0.9918

Exercise 9A

(Probabilities may be expressed as percentages or decimals.)

1 $\frac{1}{4} = 0.25$

2 0.3

3 0.1

4 (a) 0.0625   (b) 0.1875

Exercise 9B

1 (b) and (c). People are equally likely to be born on any day of the week and any of the numbers 1 to 9 are equally likely to come at the end of a telephone number, but heights of people cluster around a mean value.

2 Discrete uniform distribution

3 (a)

x:	1	2	3	4	5	6	7	8
p(x):	$\frac{1}{8}$	$\frac{1}{8}$	$\frac{1}{8}$	$\frac{1}{8}$	$\frac{1}{8}$	$\frac{1}{8}$	$\frac{1}{8}$	$\frac{1}{8}$

  (b) $\frac{1}{8}$

  (c) $\frac{1}{4}$

4 There is not an equal probability of hitting all the numbers if you aim for the 20. The distribution is not suitable.

5 (a) a discrete uniform distribution

  (b) $\frac{1}{50}$

6 $\frac{1}{5}$

7 (a) a discrete uniform distribution

  (b) $\frac{2}{6} = \frac{1}{3}$

Exercise 9C

1 (a) Binomial

  (b) 12 is number of trials, 0.325 is probability of success.

2 (a) 0.6 **(b)** 0.3456
3 (a) 0.614125 **(b)** 0.057375
4 (a) 0.15625 **(b)** 0.96875 **(c)** 0.03125
5 (a) 0.2 **(b)** 0.896
6 (a) 0.001 **(b)** 0.729
7 (a) 0.0256 **(b)** 0.2688
8 (a) 0.000 03 **(b)** 0.999
9 (a) 0.000 01 **(b)** 0.91854
10 (a) 0.614125 **(b)** 0.325125

Exercise 9D

1 (b), (c). These are continuous variables and natural occurrences so will be normally distributed. The number of accidents is not continuous.
2 (a) They are all equal.
(b) The distribution is symmetrical about the mean. 95% lie within 2 standard deviations and 99.8% lie within 3 standard deviations of the mean.
3 0.9 and 5.7 cm
4

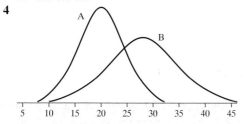

5 (a) 2
(b) (i) 14 **(ii)** 24
6 (a) 70.5 km/h **(b)** 34.5 km/h
(c) 25.5 km/h **(d)** 79.5 km/h
7 (a) 97.5% **(b)** 2.5% **(c)** 47.5%
8 (a) 95% **(b)** 47.5%
(c) 97.4% **(d)** 2.4%
9 2.4%
10 (a) 95% **(b)** 2.5% **(c)** 570
11 (a) (i) 95% **(ii)** 99.8%
(b) (i) 950 **(ii)** 998
(c) 97.4%
12 (a) 29 days **(b)** 1 day **(c)** 29 days.
13 warning 64.4 and 65.6; action 64.1 mm and 65.9 mm
14 (a) 0.025 **(b)** 0.95 **(c)** 5000 hours
15 $7\frac{1}{2}$ cm
16 25

Revision exercise 9

1 (a) 10 **(b)** 6
(c) Discrete uniform distribution
2 (a) 0.970 299 **(b)** 0.000 297
3 (a) (i) 95% **(ii)** 99.8%
(b) 97
4 (a) 47.5% = 0.475 **(b)** 97.5% = 0.975

Examination Practice Papers

Foundation Section A

1 (a) B – Becauses it is quick *or*
C – Because it is easy to see how many cars of each colour these are.
(b) A set of items selected from a population.
(c) (iii) would be best since every day is represented and every 1 hour has a chance of being chosen.
2 (a) Madonia and Elmonia.
(b) 11–30 cases.
(c) Alta since the highest number of reported cases were found here.
3 (a) 3, 5, 8, 5, Total 26. **(b)** 61–75 mins.
(c) 54.7 mins.
4 (a)

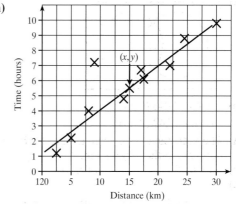

(b) see diagram. **(c)** 15.0 km
(d) see diagram. **(e)** positive correlation
5 (a) *y*-axis does not start from zero. The thick line makes it difficult to read.

Foundation Section B

1 (a) Battery hens 20
Free range hens 22
(b) 73.5 grams
(d) Free range hens 36
(e) No – lower median weight
2 (b) 58 **(c)** 15
3 (a)

B → at about 0.2   A → at about 0.5   C → at about 0.75

0   0.1   0.2   0.3   0.4   0.5   0.6   0.7   0.8   0.9   1

4 (a) $\frac{9}{64}$ **(b)** $\frac{55}{64}$
5 (a) continuous **(b)** 9.2 cm
(c) Higher median
(d) Pond A (higher interquartile range)
(e) Pond B
(f) Yes – higher median – although A does not have largest fish.

6 **(a)** 103.2
   (b) Fuel and light – price index 2003 is less than 100
   (c) £2577.50
   (d) All categories are likely to increase at same rate over such a long time period.
7 **(b)** 39, 41.7, 43, 45.7, 48.7, 50
   (c) Sales are rising since the moving averages are getting larger

Higher Section A

1 **(a)** 400 fish
   (b) Some fish may have lost marks, some died, some born.
2 **(a)** 4.24 cm
3 **(a)** A – more fish are longer
   (b) A – negatively skewed, **B** – normally distributed, **C** – positively skewed
   (c) 95%
4 **(a)** To check whether the questions are correctly worded and unambiguous. (To check whether questions are closed.)
   (b) Year 7 – 22 or 23   Year 8 – 22 or 23
       Year 9 – 18 or 19   Year 10 – 18 or 19
       Year 11 – 17 or 18
5 **(a)** Quantitative   **(b)** Qualitative
   (c) Continuous   **(d)** Discrete
6 **(a)**

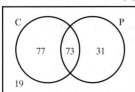

   C – computers
   P – phones

   (b) $\frac{19}{200}$
7 **(b)** $\frac{7}{18}$ **(c)** $\frac{5}{7}$

Higher Section B

1 **(b)** Yes
   (c) 0.792
   (d) There is positive corrrelation. The more points scored the more goals the team is likely to have scored.
2 **(a)** 1.455 accidents
   (b) 0.989 accidents
   (c) The earlier study showed more accidents on average and also more variability in the daily number of accidents.
3 Another sample shoiuld be taken immediately.
4 **(b)** 39, 41.7, 43, 44.7, 47.3, 48.7, 50
   (c) The trend line has a positive gradient. It suggests that there is a rising trend in sales.
5 **(a)** $y = -1500x + 12\,000$ or $y = 12\,000 - 1500x$
   (b) a is the rate of decrease in the value per year. b is the value of a new Blokeswagen Colf.
   (c) A → (ii); B → (iii); C → (i)
6 **(a)** 2 letters
   (c) 2 letters
7 **(a)** Taxes
   (b) £352
   (c) 108.48
8 **(a)** 71 mm
   (b) 13 mm
   (c) 34 mm, 47 mm, 94 mm, 100 mm
       They are all more than 1.5 × IQR from the median.
   (e) Sour tasting walnuts are longer on average and are of more consistent length.
9 **(b)** England has a much larger percentage of the population than of area.
       Scotland has a much larger percentage of the area than the population.

Index